The Institute of Biology's
Studies in Biology no. 161

Earthworm Biology

John A. Wallwork

D.Sc.

Professor of Zoology,
Westfield College,
University of London

Edward Arnold

© John A. Wallwork 1983

First published in Great Britain 1983
by Edward Arnold (Publishers) Ltd
41 Bedford Square, London WC1B 3DQ

Edward Arnold (Australia) Pty Ltd
80 Waverley Road
Caulfield East
Victoria 3145
Australia

Edward Arnold
300 North Charles Street
Baltimore
Maryland 21201
United States of America

British Library Cataloguing in Publication Data

Wallwork, J. A.
    Earthworm biology.—(Institute of Biology's
    studies in biology, ISSN 0537-9024; no. 161)
    1. Opisthopora
    I. Title   II. Series
    595.1'46   QL391.04

ISBN 0-7131-2884-4

Printed and bound in Great Britain at
The Camelot Press Ltd, Southampton

General Preface to the Series

Because it is no longer possible for one textbook to cover the whole field of biology while remaining sufficiently up to date, the Institute of Biology proposed this series so that teachers and students can learn about significant developments. The enthusiastic acceptance of 'Studies in Biology' shows that the books are providing authoritative views of biological topics.

The features of the series include the attention given to methods, the selected list of books for further reading and, wherever possible, suggestions for practical work.

Readers' comments will be welcomed by the Institute.

1983

Institute of Biology
20 Queensberry Place
London SW7 2DZ

Preface

Earthworms are familiar animals to most people, so much so in fact that many aspects of their biology are frequently taken for granted. However, there is still much to be learned about them, and this applies particularly to tropical species and those living in the southern Hemisphere. Much of the information that is available relates to British species, and the account given here reflects this fact. At the same time, earthworms from other parts of the world have not been neglected and comparisons are made wherever possible. The booklet attempts to integrate the anatomical, physiological and ecological aspects of earthworm biology in a concise and readable manner. To this end, depth has had to be sacrificed for breadth and superficiality is inevitable. Hopefully, what has emerged is a text which will stimulate students to view earthworm biology in a new light, and to encourage further research on those families and species which have been largely neglected hitherto.

London, 1983

J.A.W.

Contents

1 Annelids: the First Coelomates

1.1 Introduction

The phylum Annelida is generally considered to consist of four Classes: the Polychaeta, Oligochaeta, Archiannelida and Hirudinea, although some authorities consider that a fifth group, the Pogonophora, warrants inclusion here (see CLARK, 1978). Polychaetes are almost exclusively marine in habit, and frequently have parapodia: segmentally-arranged extensions of the lateral body wall shaped like paddles that variously help in swimming, food collection and, in the case of some tube-dwelling or sedentary forms, ventilation. As the name suggests, the members of this Class are equipped with numerous chaetae on each body segment. The life cycle also includes a trochophore larva; a ciliated, motile form shaped rather like a spinning top, that is planktonic or benthic in habit.

The Archiannelida is a heterogeneous group of primitive and secondarily reduced forms, with chaetae reduced or absent, and they are indistinctly segmented. In keeping with their simple organization, these annelids often have a reduced coelom and vascular system. A larval stage is present in some of these archiannelids but not in others.

The Hirudinea, more commonly known as leeches, are examples of specialized or regressive evolution; they have adopted a carnivorous or blood-sucking ectoparasitic mode of life and they are equipped with at least one, usually two, suckers to facilitate locomotion and attachment to their host. The coelom is reduced and chaetae are almost always absent. Embryonic development is direct, without the intervention of a distinct larval stage in the life cycle.

The Pogonophora are marine tube-dwelling worms in which the body is divided into four regions: the pro-, meso-, meta- and opisthosome – each with its own coelomic compartment. They do not possess a digestive system but the segmentation of the opisthosome is very reminiscent of that found in the annelids generally.

The Oligochaeta, to which earthworms belong, occur mainly in freshwater and terrestrial habitats. The segmentally-arranged chaetae show a reduction from the polychaete condition to four pairs, as a general rule, but they may be more numerous in some groups (which are therefore described as perichaetine). They lack parapodia and suckers, and development is direct without the intervention of a free-swimming larval stage in the life cycle. The freshwater oligochaetes can be distinguished, in general, from their terrestrial relatives by the degree of development of their excretory organs, the nephridia. These organs are tubular in form and they open internally as a ciliated funnel, or nephrostome (see Fig. 5–2). They convey excretory materials that accumulate in the coelomic fluid to the exterior. The upper lip of this funnel is well-developed in earthworms, and reduced to a few marginal cells, or absent, in aquatic forms. This is a fine anatomical distinction that may escape the attention of beginners, but it presents no insuperable problems of identification as far as the treatment in this book is concerned. Earthworms are annelids that live in soil and decaying organic matter, and are generally recognizable as such.

1.2   The coelom

The advent of the annelids into the invertebrate evolutionary story provided a fundamental step forward in body design. This anatomical advance was signalled by the appearance of the coelom, a space surrounding the gut and separating it from the body wall. The annelid coelom is formed during embryonic development when sheets of mesoderm cells appear in the blastocoele (the primary segmentation cavity) and eventually separate into outer and inner layers (somatic and splanchnic, respectively) as the intervening cells migrate or disintegrate (Fig. 1–1). The cavity so formed largely or

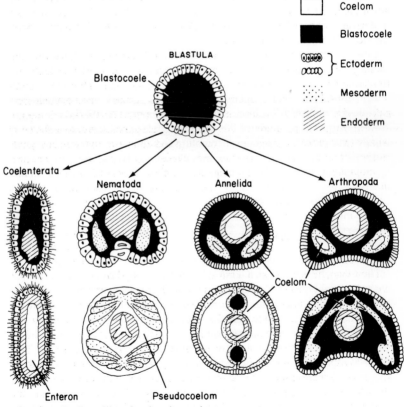

Fig. 1–1   Body cavities of various invertebrates.

completely obliterates the blastocoele, and is filled with fluid containing amoeboid cells. The sheets of somatic and splanchnic mesoderm differentiate into muscles of the body wall and gut, respectively, and into epithelium, or peritoneum, which envelops the gut and other viscera.

It is by no means certain why a coelom developed in the first place. Its original function, which it frequently retains today, may have been for the accumulation of gametes and/or excretory products that are voided via coelomoducts and nephridia. On the other hand, the separation of the body wall musculature from that of the gut was a necessary event, if locomotory activities controlled by muscles of the body wall were to proceed independently of the activities of gut muscles controlling the passage of food down the alimentary canal. The separation of these two functions would increase the potentialities for both more efficient locomotion and a more efficient digestive system.

1.3 Body cavities

Whatever its origins, the presence of a coelom confers considerable advantages on annelids in general, and the oligochaetes in particular. To understand why this is so, it is necessary to outline the development of body cavities in more lowly metazoans.

Coelenterates, primitive Metazoa in which the basic body design is that of a radially symmetrical blind sac, have but a single body cavity, the coelenteron, that connects to the exterior *via* a single opening, functioning both as a mouth and an anus. Provided this aperture is tightly closed, contraction of the muscle cells in the body wall causes a pressure to be exerted on the fluid filling the coelenteron. This pressure then acts on other, relaxed muscle cells in the body wall, causing them to lengthen. In this way, a deformable body can change its shape and move. The familiar looping movements of *Hydra* exemplify this type of movement. But the disadvantage here is that the animal cannot feed and move at the same time because fluid pressure in the coelenteron will be dissipated if the mouth is opened.

1.4 Hydrostatic skeletons

We can readily appreciate that a fluid-filled body cavity, such as a coelenteron, functions as a kind of skeleton; a hydrostatic skeleton that controls the body shape and acts as an agent of muscular antagonism. To use a body cavity, that also serves as a digestive cavity, for these purposes as well is clearly not an ideal arrangement. The Platyhelminthes show an improvement over the coelenterate design in that the musculature of the body wall is separated from that of the gut by mesenchyme, or parenchyma. But the only body cavity is the gut, and it has a single opening. The nemertine worms (Nemertea) represent an advance on this condition with the development of a through gut, equipped with mouth and anus. However they, like the Platyhelminthes, cannot use this body cavity as a hydrostatic skeleton to any extent – contractions of the body wall muscles are considerably dampened by the mesenchyme and have little impact on the fluid in the gut cavity. Clearly, any improvement on this design must involve the development of a fluid-filled body cavity that is independent of the gut and on which the muscles of the body wall can act directly. The nematodes and their relatives have such a cavity, derived from vacuolated mesoderm cells and termed a pseudocoelom. This serves adequately to maintain body shape, but it is of limited use for locomotion since, in the nematodes at least, a thick cuticle severely restricts any change in the shape of the body.

The problem has been solved much more effectively by the annelids in which a secondary body cavity has developed within the mesoderm, a true coelom. This, together with an elastic body wall, provides an antagonistic system in which contractions of the body wall musculature can bring about changes in body shape without affecting the normal functioning of the gut.

1.5   Peristaltic locomotion

The body wall of an annelid such as an earthworm is equipped with both longitudinal and circular muscles. When circular muscles contract at the anterior end of the body, the coelomic fluid in this part is subjected to radial compression. Since the volume of the coelomic fluid is constant, the fluid pressure generated will have the effect of lengthening the longitudinal muscles, which are in a relaxed state. The anterior end of the worm, narrowing in diameter, will extend forwards. Following this extension, the longitudinal muscles contract, causing a longitudinal compression of the coelomic fluid and, as the circular muscles relax, the body will bulge at the anterior end. Successive waves of contraction and relaxation, of constriction and bulging, produce a peristaltic type of movement. But this would not be possible without one more, very important, innovation.

1.6   Septa

In an animal where the coelom is continuous throughout the body (and there are such, e.g. worms belonging to the phylum Sipunculida), the pressure generated on the coelomic fluid in any one part of the body is quickly transmitted down the fluid column to every other part. Peristaltic locomotion is characterized by successive waves of circular and longitudinal compression. The radial and longitudinal pressures associated with this type of locomotion could interfere with each other in a continuous coelom. In order to separate these two pressure waves, some form of compartmentalization of the coelom is required. This is provided in the annelids by the development of septa; muscular partitions that divide the body into segments.

Among the annelids, the development of septa is best seen in the earthworms. These animals have perfected a type of locomotion in which the body moves forwards by protrusion as successive peristaltic waves of muscular contraction pass from anterior to posterior. This locomotory pattern is ideally suited to the worm's habit of burrowing in soil and organic matter. The initial lengthening and narrowing of the anterior end as the circular muscles contract allow this part of the body to penetrate small crevices in the soil. The peristaltic bulge that follows serves to push soil particles aside and create a burrow. Earthworms are circular in cross-section, and their cylindrical body shape enables them to exert a considerable thrust against the wall of the burrow with the entire circumference of their body wall musculature. Forward movement is facilitated by the presence of chaetae that can be extended to anchor one part of the body while another is being drawn forwards. They can also be withdrawn or turned on their axis to facilitate bending movements. In compacted soils, where peristaltic movement may have little effect in creating a burrow, earthworms remove soil particles in their path by eating them.

1.7   Metamerism

The compartmentalization of the annelid body that results from the development of septal partitions has important repercussions on the arrangement of the internal organs. Septa effectively isolate each body segment which must, then, become virtually self-sufficient. Each segment must have its own supply of blood, its own central and peripheral nervous system, its own excretory system for removing metabolic wastes, and its own gonads. As a consequence, blood vessels, ganglia and nerves, excretory organs (nephridia) and gonads may be present in each segment. This serial repetition of organs throughout the annelid body is termed metamerism. Earthworms illustrate some variations on this metameric theme, as we will see later.

1.8   Dorsal pores

Although not conspicuous in most earthworms, dorsal pores are present on the inter-segmental region throughout much of the body length (they are lacking on a few of the most anterior segments). These pores are the avenues for the discharge of coelomic fluid to the exterior. This discharge may serve as a repugnatorial secretion, or a means of keeping the body surface moist to facilitate respiration and prevent desiccation.

2 Earthworm Families and their Distribution

2.1 Introduction

The term 'earthworm' has been used in different ways by different authors. In Great Britain, 'earthworms' belong, almost exclusively, to the family Lumbricidae, although another family of oligochaetes with terrestrial representatives, the Enchytraeidae, is equally common if less conspicuous. In West Africa, many of the common earthworms belong to the family Eudrilidae; in South Africa there is the Microchaetidae, in Australia and other parts of the Far East, the Megascolecidae, and so on. Perhaps it would not be inappropriate at this stage to document the families of earthworms, and to outline their distribution patterns on a global scale.

2.2 Families of earthworms

The classification of earthworms at the family level is the subject of some controversy, although this need not concern us unduly here (see the discussion in EDWARDS and LOFTY, 1977). Essentially, the truly terrestrial earthworms of the world can be assigned to no more than a dozen families (the exact number depends on who is doing the classifying). This is a remarkably small number of taxa for a group of invertebrates that has been around for a very long time, and that has been so successful in spreading to soils throughout the world. There is a sharp contrast here, for example, with soil-dwelling saprophagous mites that occupy a similar ecological niche. These mites are represented by families well in excess of a hundred world-wide. The variety of substrates available to these microarthropods is equally available to earthworms, and this should not present any obstacle to their diversification. However, inter-group competition for available food (see Chap. 9) could be a potent factor in limiting evolutionary diversification. This point is illustrated by the Lumbricidae, a family that has a distribution predominantly in the Northern Hemisphere. When introduced into Southern Hemisphere localities, lumbricids compete successfully with the indigenous earthworm fauna to the extent that they exclude them (with, perhaps, the exception of species of the megascolecid *Pheretima*). However, attractive though this line of reasoning may appear, it does not hold up under detailed scrutiny. Earthworms do show some feeding specificity (see Chap. 3) and, by this token, should be able to exploit a variety of food niches and refugia. Apparently they have not been able to do this in an evolutionary sense.

Two groups of earthworms predominate as far as numbers of species and individuals are concerned: the megascolecids and the lumbricids. The former consist of four families: Megascolecidae, Acanthodrilidae, Octochaetidae and Ocnerodrilidae, although the last three have been relegated to sub-family status in a review by JAMIESON (1970). The Megascolecidae (sensu stricto) contains 20 genera, and included in this list are *Digaster*, *Lampito* and *Pheretima* mentioned elsewhere in this book. There are 24 genera in the Acanthodrilidae, while the Octochaetidae comprise 36 genera, and the Ocnerodrilidae 21 genera.

The lumbricid grouping consists of one family, the Lumbricidae, with 10 genera, and it is perhaps the most widespread of all the earthworm families. The remaining families are the primitive Moniligastridae (5 genera), Eudrilidae (40 genera), Glossoscolecidae (16 genera), Microchaetidae (6 genera) and the monogeneric Sparganophilidae, Hormogastridae and Criodrilidae.

2.3 Earthworm biogeography

Two families of earthworms appear to be indigenous to the Northern Hemisphere: the Lumbricidae and Criodrilidae. The Hormogastridae are principally Mediterranean, while the remaining 9 families have a distribution that is centred in the Southern Hemisphere. These biogeographical divisions are, to some extent, obscured by the presence of introduced (peregrine) species, notably those belonging to the family Lumbricidae, that have become established alongside native species. In biogeographical studies, these introduced species must be clearly distinguished from the native, or endemic, species.

Although it has its detractors (see STEPHENSON, 1930), the view is widely held that the oligochaetes are an ancient group. They probably arose before the main continental masses moved apart (during the Triassic) to assume their present positions. Given this, the patterns of biogeographical distribution observable today may provide clues as to the evolution of the various earthworm groups. If we look, for example, at the Lumbricidae and disregard the peregrine species that have become world-wide in distribution, we can find no evidence that this family evolved early enough in geological time to have allowed its dispersal to all of the major continental masses before they drifted apart. The implications here are two-fold: namely that the oligochaetes originated in the Northern Hemisphere, and that the Lumbricidae are one of the most recent products of this evolution. According to this line of reasoning, families such as the Moniligastridae of south-east Asia, the Megascolecidae distributed across much of the Southern Hemisphere, and the Eudrilidae of Africa would be considered 'primitive'. Their early advent on the evolutionary scene would allow them to spread into South America, Africa and the Far East before the southern continents separated and drifted to their present positions. Subsequently, the argument goes, these families or their progenitors became extinct in the Northern Hemisphere, unable to compete with the lately evolved Lumbricidae.

The fossil record of the oligochaetes is poorly documented, and we can find little support in this for the above arguments. On the other hand, some support is provided by studies on the comparative anatomy of the groups occurring today. The structural organization of the Moniligastridae is very reminiscent of the primitive aquatic Lumbriculidae. The Megascolecidae and the Eudrilidae represent some evolutionary advance but are considered more primitive than the Lumbricidae in some respects. For example, sperm transfer during copulation is direct – by direct apposition of male pores to the spermathecal openings of the partner – whereas this is not the case in lumbricids. The latter possess a more intricate mechanism for sperm transfer (see Chap. 7 and EDWARDS and LOFTY, 1977, for further discussion).

The possibility that lumbricids, introduced accidentally by human traffic into Southern Hemisphere localities, have competed successfully against native species deserves further attention. There are differing opinions on this point. Advocates of this view argue that the success of lumbricids in the Northern Hemisphere is an indication of their colonizing ability in a range of soil types and microclimates. This ability would confer a distinct advantage when they were introduced into new environments. Against this, there is no direct evidence that lumbricids actually displace native species (LJUNGSTRÖM, 1972). Megascolecids, microchaetids and eudrilids cast at the surface which suggests that they are effective burrowers. Competition among subterranean animals can often be intense (see Chap. 9), and this raises the possibility that these rather primitive groups may have to enter into competitive situations where they co-exist naturally, and with lumbricids that may be introduced accidentally.

In such cases, lumbricids may have a competitive advantage by virtue of their extended periods of activity (tropical earthworms usually enter into obligatory diapause – quiescence – at certain times of the year (Chap. 7)). Further, even the deep-burrowing species of lumbricids appear on the surface at frequent intervals to feed on the accumulated nitrogen-rich leaf litter. Megascolecids and microchaetids, which burrow to greater depths than lumbricids, may visit the soil surface only sporadically and could be at a disadvantage when it comes to the exploitation of nitrogen 'pools' present in surface accumulations of leaf litter. These are theoretical possibilities, but they require empirical confirmation.

These arguments focus attention on one critical aspect of earthworm biology – the ability to burrow that is expressed to varying degrees by different species. We can conclude this chapter by exploring this theme in a little more detail.

2.4 Life styles and the burrowing habit

As just noted, some earthworms produce very deep burrows extending several metres down the soil profile. The giant megascolecids of Australia and the microchaetids of South Africa illustrate this point very well. In Britain however, only *Lumbricus terrestris*, *Allolobophora nocturna*, *A. longa* and *Octolasion cyaneum* burrow to any depth (1m or more). Other species may be restricted to horizons in the soil profile nearer the surface. The common *Allolobophora caliginosa* and *A. chlorotica*, for example, produce shallow burrows, whereas *Lumbricus rubellus* does not really burrow at all – it lives in decaying organic material that has accumulated at, or near, the soil surface.

Clearly, different species of earthworms have different life styles and these have been formally classified into the following categories by BOUCHÉ (1977):

(*1*)   *Épigés*   Essentially litter dwellers; small in body size (between 10 and 30 mm usually); uniform colouration. Examples include *Lumbricus rubellus*, *L. castaneus*, *Dendrobaena octaedra*, *D. mammalis*, *D. rubida* and *Eisenia foetida*.

(*2*)   *Endogés*   Species such as *Allolobophora caliginosa*, *A. rosea*, *A. muldali* and possibly *Octolasion cyaneum* that construct horizontal, branching burrows in the organo-mineral layer of the soil. The members of this category are weakly pigmented, and vary in size from small to large.

(*3*)   *Anéciques*   Deep-burrowing species that construct vertical burrows, cast at the surface and emerge from their burrows at night to draw down organic material. British examples include *Lumbricus terrestris* and *Allolobophora longa*. Characteristically, anéciques are large in size as adults (200 to 1100 mm long); anteriorly and dorsally they are dark brown in colour.

These three categories have other distinctive attributes and these will be introduced in the appropriate places later in this book. It is sufficient to note, at this stage, that these divisions are not absolute; some species cannot be so neatly assigned and in agricultural soils, for example, earthworms in general burrow to greater depths than they do in more compacted grassland and forest soils. Again, these categories can be sub-divided. The endogé group, for example, can be split into 'epiendogé' and 'hypoendogé' depending on whether the species in question select the upper or the lower part of the organo-mineral horizon.

3   Food, Feeding and Casting

3.1   Introduction

Earthworms are, for the most part, saprophages. They feed mainly on organic detritus, usually the decomposing leaves and stems of plants, although root material, seeds, algae, fungi and testacean Protozoa may also be ingested (PIEARCE, 1978). This might suggest that earthworms will feed on anything they can find in their environment that is organic in nature. However, many of the common British species, such as *Lumbricus terrestris, Allolobophora caliginosa, A. longa, A. rosea* and *Dendrobaena octaedra*, are known to prefer food materials rich in nitrogen and sugar and low in polyphenols (SATCHELL, 1967). These include the tissues of herbaceous plants (e.g. nettle and dog's mercury) and the leaves of some tree species such as ash, elm, alder, birch and sycamore. Conifer needles and the leaves of oak and beech, with their higher carbon/nitrogen ratios and polyphenol contents are less acceptable.

Variable amounts of mineral soil may be ingested along with this organic material. Litter-dwelling épigés, such as *Lumbricus rubellus, L. castaneus* and *Dendrobaena* spp., together with the deeper-dwelling endogés, *Allolobophora caliginosa, A. rosea* and *Octolasion cyaneum*, eat their way through a substratum that contains organic and mineral particles in an advanced state of decomposition. Their gut contents include a mineral fraction which reflects that of the external medium. In forest soils, this fraction is minimal, but it increases appreciably in the soils of permanent pastures. Deep-burrowing anéciques emerge at night and draw down fragments of decaying leaves into their burrows; here they are stored until sufficiently decomposed to be eaten. The amount of mineral soil present in the gut contents of these worms will vary with their burrowing activity. When they are engaged in excavating a burrow, they ingest large quantities of mineral soil, and this is eventually voided at the surface as a cast. At other times, organic fragments predominate in the gut contents.

BOUCHÉ (1977) has used the terms 'microphage', 'mesophage' and 'macrophage' to distinguish between endogés, épigés and anéciques, respectively. This distinction is based on the size of the ingested particles, ranked as small, medium or large. It has been found useful as a way of identifying how organic resources are partitioned between different, co-existing species of earthworms.

In an ecological sense, saprophages are primary consumers – they provide a pathway for energy-rich organic material, produced by the photosynthetic activity of green plants, to be converted into animal cytoplasm, and thereafter to be utilized for maintenance, growth and reproduction. It is not our purpose, at this point, to investigate these aspects of trophic activity, but it is pertinent to register the fact that the passage of food through the gut varies with the ecological group. It is probably slower in épigés than endogés, which may reflect the fact that the food of the latter is in a more advanced stage of decay than the food of the former, and is therefore digested more rapidly. In anéciques, such as *Lumbricus terrestris, Allolobophora longa* and *Octolasion* sp., gut transit time is variable. The members of this grouping consume between 10% and 30% of their live body weight per day.

3.2   The alimentary canal

The earthworm gut is a straight tube running from the mouth to the anus, and it shows regional differentiation. Anteriorly, the mouth, which is flanked dorsally by a lobe-like flap, the prostomium, leads to a muscular pharynx which pumps food backwards into the next part of the digestive tract, the oesophagus. The pharynx is equipped with glands, secreting mucus to lubricate its walls and facilitate the passage of food material. Associated with the oesophagus are calciferous glands that are variably developed in different species of lumbricids. In their simplest form, these glands are merely foldings of the oesophageal wall, and are virtually inactive as far as calcium metabolism is concerned. In a more advanced state of development, the glands assume their own identity and connect with the lumen of the oesophagus *via* a duct; examples of this type occur in *Lumbricus castaneus, L. rubellus* (Fig. 3–1c) and *Dendrobaena rubida subrubicunda*. Originally it was suggested that the main function of these glands was to excrete excess calcium, in the form of calcium carbonate, from the body tissues into the gut and thence to the exterior through the anus. Many earthworms live in soils rich in calcium and it would seem appropriate for them to have a mechanism for ridding the body of excessive amounts of this element. However, this is not the entire story for PIEARCE (1972) reported that *Dendrobaena veneta hibernica typica*, which lives in calcareous woodland soil, passes considerable quantities of calcium through the alimentary canal, but that its calciferous glands are of the simplest type (Fig. 3–1a) and apparently are inactive. Recently it has been demonstrated that active calciferous glands, in some species, contain large quantities of carbonic anhydrase, an enzyme that figures prominently in acid-base reactions. Metabolic activities produce carbon dioxide that can increase the acidity of the coelomic fluid. Carbonic anhydrase in the calciferous glands apparently counteracts this effect by catalysing the fixation of this carbon dioxide in the form of calcium carbonate. The glands are, in part at least, regulators of the pH of the coelomic fluid, and experimental removal of these glands results in a lowering of the coelomic pH.

Immediately behind the oesophagus the gut is usually differentiated into two bulbous chambers, the crop and gizzard respectively; these, like the oesophagus are lined with cuticle. The thin-walled crop stores food material prior to its passage into the gizzard – it may be considered as a collecting station. The thick, muscular walls of the gizzard grind the food into small

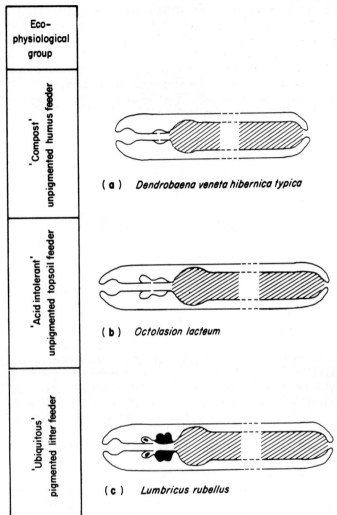

Eco-physiological group	
'Compost' unpigmented humus feeder	(a) *Dendrobaena veneta hibernica typica*
'Acid intolerant' unpigmented topsoil feeder	(b) *Octolasion lacteum*
'Ubiquitous' pigmented litter feeder	(c) *Lumbricus rubellus*

Fig. 3–1 Ecophysiological groups of British lumbricids represented to show the varying development of the calciferous glands. (a) Simplest type, inactive; (b) intermediate type, active; (c) advanced type, highly active.

particles which then pass into the posterior part of the alimentary canal, the intestine. This type of arrangement is common in lumbricids (Fig. 3–2), but in the megascolecids the crop and gizzard (when present) are situated immediately behind the pharynx. Megascolecids are very variable as far as the development of the gizzard is concerned. This part of the gut is lacking in some species, but in others there may be as many as ten gizzards present, segmentally arranged.

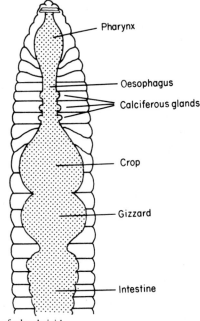

Pharynx

Oesophagus

Calciferous glands

Crop

Gizzard

Intestine

Fig. 3–2   The digestive system of a lumbricid.

The intestine of most earthworms that have been studied is remarkable among annelids in being infolded in the median dorsal line to form a typhlosole. This infolding effectively increases the internal surface area of the intestine and, thereby, the secretory and absorptive surface area of this region of the gut. Such a device is common among animals whose principal diet is plant material, since this material requires a prolonged period of digestion, although the means by which this additional surface area is achieved varies in different groups of saprophages and herbivores. In lagomorph mammals (rabbits and hares), for example, the appendix is greatly developed; in ruminant mammals, the stomach consists of three or four chambers; earthworms typically have a typhlosole. The typhlosole of *Lumbricus terrestris* is present along the anterior two-thirds of the intestine, but is lacking along the posterior position. In *Megascolex* and *Eiseniella*, the typhlosole is little more than a ridge hardly projecting into the intestinal lumen; in *Megascolides australis* there is no typhlosole at all.

Food material passes down the gut mainly by the peristaltic action of muscles in its wall. However, parts of the gut epithelium in which a cuticle is thin or lacking, are ciliated, and the action of these cilia may supplement that of the gut musculature. Ciliated epithelium is present in the posterior part of the oesophagus where it may well function to facilitate the passage of food into the mid-gut. Ciliated cells are also present on the typhlosole, but apparently these do not aid the passage of food (MICHEL and DEVILLEZ, 1978).

3.3 Digestive enzymes

A variety of enzymes has been reported from the alimentary canal of earthworms and it would be unwise to generalize. Experience with other soil saprophages, such as oribatid mites, indicates that the complement of digestive enzymes is closely related to the preferred diet. For example, many soil saprophages that feed on fungi possess trehalase. The disaccharide trehalose occurs commonly in fungi but is rare in other plant material; earthworms feed primarily on the latter and apparently do not have trehalase. Again, members of the genera *Lumbricus* and *Allolobophora* secrete a proteolytic enzyme from the pharyngeal glands. No proteolytic activity has been detected in the pharynx of *Eisenia foetida* however, although an amylase is present (van GANSEN, 1962).

The intestine is the main region for digestion and absorption. Enzymes recorded from here include proteases, a lipase, amylase, cellulase, lichenase and chitinase (see LAVERACK, 1963). These operate in a milieu that has a remarkably stable pH, varying only between 6.3 and 7.3 throughout the length of the intestine. All of the available evidence points to the lumen of the intestine as the focal point of digestive activity in earthworms. Intracellular digestion (phagocytosis) does not seem to be very important in these animals.

3.4 Source of digestive enzymes

Earthworms, like many other soil-dwelling saprophages harbour gut symbionts: bacteria, actinomycetes and protozoans. That the earthworm gut provides a suitable environment for the growth of bacterial colonies is evidenced by the fact that earthworm castings contain significantly higher counts of bacteria than does the surrounding soil. These symbionts are capable of secreting digestive enzymes and the question is frequently asked: do the enzymes in the earthworm gut originate from the gut epithelium or from these symbionts? This is largely an academic question, for it is the fact that these enzymes are there that is important, not where they came from. It is also a fact that enzyme-secreting cells occur in the epithelium lining the intestine, and it is unlikely that earthworms need to rely on gut symbionts for their supply of digestive enzymes. Why then – and perhaps this is a more pertinent question – do earthworms have such symbionts?

A symbiosis between an earthworm and its gut microflora is biologically interesting for it focusses attention on the factors that allow such a harmonious and beneficial relationship to develop and be sustained. One such factor may be the relatively stable and high pH of the intestinal contents; such conditions encourage the growth of bacterial colonies. Recently, MARIALIGETI (1979) demonstrated that members of the genus *Vibrio* occurred in enormous numbers in the posterior part of the gut of *Eisenia lucens*, an earthworm that tunnels in decomposing beech stumps in Hungary. Members of this bacterial genus constituted 73% of 473 strains isolated from the gut contents. These rod-shaped, Gram-negative, facultative anaerobes showed a wide range of biochemical capabilities including the ability to convert nitrate to nitrite. They can also break down dextrose, arabinose, xylose and, to a lesser extent, lactose and mannitol. However, tests carried out to determine their ability to degrade cellulose and lignin proved negative.

This same species of earthworm also has an intestinal actinomycete flora comprising, for the most part, a single species, *Streptomyces lipmanii* (CONTRERAS, 1980). The numerical dominance of this particular species of microorganism, which is not so abundant in other milieux, suggests that the earthworm gut exerts a selective influence on *S. lipmanii*, favouring its development here. It is possible that actinomycetes such as *S. lipmanii* participate in the chemical transformation of organic materials, the formation of clay/humus complexes, and the production of cementing substances that improve the crumb structure of the soil. If indeed this proves to be the case, then gut communities of earthworms have an important place in the decomposer system of the soil.

3.5   Earthworms and soil formation

The soil-forming process is, essentially, a process of decomposition. The products of this decomposition, in a fertile soil, originate from two sources. Firstly, the mineral parent material, be it bedrock or a transported deposit, undergoes physical, chemical and biological weathering. The second source is organic material that originates from the dead parts of aerial vegetation that fall on to the surface of the soil and from an input derived from the *in situ* decomposition of plant roots. The decomposition process itself – and this applies both to the mineral and organic material – incorporates physical and chemical processes. Physical decomposition involves the progressive comminution of larger particles into smaller ones. Chemical decomposition results in the degradation of complex molecules into simpler ones. These two events go hand in hand, for the physical decomposition exposes a greater surface area of soil particle on which chemical changes can take place. Ultimately, in a fertile soil, the products of organic and mineral decomposition become intimately associated as organo-mineral, or clay/humus, complexes. These provide sites for the adsorption of cations, thus preventing their loss, through leaching, from the soil.

Earthworms may participate in this soil-forming process in 5 ways:
(1)   through their influence on soil pH;
(2)   as agents of physical decomposition;
(3)   by promoting humus formation;
(4)   by improving soil texture;
(5)   by enriching the soil.

3.5.1   Influence on soil pH

As already noted, the pH of the intestinal contents is remarkably stable around the neutral to slightly alkaline level. This is probably not due to secretions from the calciferous glands, as was previously thought, since these secretions are in the form of resistant calcite granules. A more probable neutralizing factor is the ammonia secreted into the intestine. The production of large amounts of alkaline faecal material can have a profound effect on the overall level of soil pH and on the course of organic decomposition. In neutral or slightly alkaline conditions bacterial activity is favoured. These microorganisms effect a more complete breakdown of organic compounds, producing a mull-type humus, than do soil fungi that flourish under more acidic conditions, and produce a mor-type humus. The role of the earthworms in promoting these conditions in nature is poorly understood. Further research may profitably be directed towards studying the commercial implications of improving the quality of acidic soils through the addition of earthworm 'manure'.

3.5.2   Physical decomposition

The passage of organic material through the earthworm gut undoubtedly results in the physical disintegration of this material. The muscular action of the gizzard (and probably also the crop) is particularly important in this respect. The grinding action of this part of the gut may be enhanced by the ingestion of silica granules. These will have an abrasive effect on the softer material of plant origin that forms the bulk of the diet. The cycling of silicon is especially common in grassland plant/soil systems and it is here, rather than in acid woodland soils, that earthworms are in high densities. However, apart from the work of PIEARCE (1978), there are few recent investigations on the composition of the gut contents of lumbricids or any of the other families for that matter. Piearce reported that mineral particles were abundant in the ingesta of six species of lumbricids (*Allolobophora caliginosa*, *A. chlorotica*, *A. longa*, *Dendrobaena mammalis*, *Lumbricus castaneus* and *L. rubellus*) taken from permanent pasture in North Wales. He found that the average size of the particles was significantly higher in the two largest species, *A. longa* and *L. rubellus*, than in the remainder. This finding is not entirely in accord with BOUCHÉ'S (1977) postulate mentioned earlier (p. 10), namely that the épigé *L. rubellus* should be mesophagous, while the anécique *A. longa* should be macrophagous. However, it does suggest that particle size selection may occur in various earthworm species. This could provide a mechanism for allowing the co-existence of several different species within the same general locality, i.e. provide a partial explanation for observed levels of earthworm species diversity.

3.5.3   Humus formation

The process of humus formation is often characterized by the selective breakdown of cellulose. The end product is a complex mixture of various organic acids, amino acids, polyphenols and sugars such as glucose, galactose, mannose, arabinose and xylose. Lignin fibres are present in 'raw' humus and peat but are degraded to polyphenols in well-decomposed humus.

The presence of cellulase in the intestine of at least some earthworms suggests that these animals may play an active role in humus formation, but there is very little hard evidence to substantiate this. As already noted, bacterial symbionts in the posterior part of the gut of *Eisenia lucens* apparently do not produce a cellulase, so that the presence of this enzyme may not be a general feature of earthworms – more research is needed in this area. Further, the structural cellulose in the walls of plant cells is usually encrusted with lignin and, unless this is removed, the cellulose will not be susceptible to enzymatic attack. Ligninase has been demonstrated in a few soil animals, notably slugs, but not in earthworms. The ability of the latter to participate, to any significant extent, in the formation of humus must, therefore, be called into question.

3.5.4 Improvement of soil texture

It has long been established that earthworms improve the 'quality' of a soil by their presence (DARWIN, 1881). But what, exactly, does this improvement process involve? The physical comminution of organic particles, the amelioration of soil pH, the enhancement of microbial (decomposer) activity are, as we have seen, all results of earthworm activity that contribute to soil fertility. All of these effects are re-inforced by mixing of the soil from different strata in the profile. Burrowing species are instrumental in this mixing process and they act, in this respect, at two levels. *Firstly*, by ingesting a mixture of organic and mineral particles, they promote the formation of organo-mineral complexes. These complexes are formed in various ways, notably through the agency of organic and inorganic cements. Electrostatic bonding may also occur between negatively charged organic particles and cations, such as calcium. Organo-mineral 'crumbs' may be formed in this way, and these improve the texture, or tilth, of the soil. Additionally, these complexes incorporate a pool of metallic ions that are held in the soil, and are not lost by leaching. Crumb formation is also promoted by the secretion of a thin, translucent peritrophic membrane by the anterior part of the typhlosole. This provides an envelope within which faecal particles are packaged before being discharged from the body as casts. This discrete packaging of soil material improves soil porosity by increasing the diameter of soil spaces, thereby improving the aeration and drainage qualities – further enhanced by the creation of burrow systems.

Secondly, by casting at the surface, earthworms bring organo-mineral crumbs from the deeper parts of the profile to the surface. Deep-burrowing species may also draw fragments of organic material (sometimes entire leaves!) from the soil surface into their burrows in the mineral soil. This two-way interchange of organic and mineral material prevents the accumulation of a surface layer of acid humus, and promotes the dispersion of finely decomposed mull humus down the profile.

Earthworms that burrow deeply into the mineral strata and return, periodically, to cast faecal material at the soil surface may facilitate the transport of certain elements to the surface litter from deep in the profile. There is abundant evidence that concentrations of exchangeable calcium, sodium, magnesium, potassium and available phosphorus and molybdenum are higher in earthworm casts than in the surrounding soil. This appears to be a general phenomenon for it has been reported from U.S.A., New Zealand, Europe, Africa, India and elsewhere. WATANABE (1975) for example, reported on the casting activity of *Pheretima hupeiensis*, a megascolecid indigenous to South

China, Korea and Japan. He found that differences existed in the chemical properties of surface casts compared with the surrounding soil in a grassland site dominated by clover (*Trifolium repens*), sorrel (*Rumex acetosa*) and plantain (*Plantago asiatica*). pH and concentrations of carbon, nitrogen, calcium and magnesium were greater in earthworm casts than in soil. This study indicates that earthworms such as *P. hupeiensis* have a positive effect on the circulation of plant nutrients.

Not all earthworms produce casts, of course, and those that do cast discontinuously throughout the year. In Britain, only *Allolobophora longa, A. nocturna* and, to a lesser extent, *Lumbricus terrestris* produce definite surface casts, and these mainly during the spring and autumn (MADGE, 1969). In Japan, *Pheretima hupeiensis* casts from mid-April to the end of October, whereas the lumbricid *Allolobophora japonica* does not produce casts at all (WATANABE, 1975). The Nigerian eudrilids studied by MADGE (1969), *Hyperiodrilus africanus* and *Eudrilus eugeniae*, produce surface casts, but only during the wet season (May to October); this pattern seems to be repeated throughout the tropics and sub-tropics.

Earthworm casts vary in size and shape and are often typical of the species producing them. In Britain, *Allolobophora longa* produces small, mound-shaped casts, but tropical species may deposit much more conspicuous casts. These may take the form of towers (*Hyperiodrilus africanus* in Nigeria and *Notoscolex* spp. in Burma), chimneys (*Dichogaster* sp. in Africa) or pyramids (*Eudrilus eugeniae* in Nigeria). Microchaetids in South Africa deposit their tower-like casts on the side of round or elliptical 'cups', called *kommetjies*, that may be up to 1 m in diameter and 30–100 cm deep (LJUNGSTRÖM and REINECKE, 1969).

The amount of soil brought to the surface by casting is quite considerable, as the data in Table 3–1 show. Some idea of the rate of soil mixing can be obtained when it is realised that the estimates given by DARWIN (1881) represent the annual deposition of a surface layer about 5 mm thick. It is evident that much higher rates of mixing occur in the tropics and sub-tropics.

Table 3–1   A comparison of surface casting activity of different earthworm faunas.

Habitat	Family	Mean wt of casts $(kg^{-1} acre^{-1} yr^{-1})$
English pasture	Lumbricidae[1]	7500–16000
West African grassland	Eudrilidae[2]	70000
South African grassland	Microchaetidae[3]	20000–100000

[1]DARWIN (1881). [2]After MADGE (1969). [3]LJUNGSTRÖM and REINECKE (1969).

3.5.5   *Soil enrichment*

The soil mixing process just described facilitates the movement of ions up and down the soil profile. It should be distinguished clearly from other processes that render ions and other nutrients more readily available to plant rooting systems. These processes involve chemical changes that release potential nutrients from 'bound', and unavailable, states into forms that can be assimilated by plants. Earthworms may participate in these processes directly and indirectly.

An example of a direct effect of earthworms is the 'mineralization' of nitrogen. This element is commonly bound in organic complexes and, as such, is not readily available to plants. Passage through the earthworm gut apparently converts this bound nitrogen into more readily 'available' forms, such as ammonia, nitrites and nitrates. Considerable quantities of nitrogen, locked up in earthworm tissues when these animals are alive, will also be released and made available on their death.

As an example of the way in which earthworms can participate indirectly in soil enrichment, the relationship between these soil animals, microorganisms and vitamin concentrations can be cited. Apparently, concentrations of vitamin B_{12} increase in soil in the presence of earthworms. ATLAVINIYTE and DACIULYTE (1969) monitored the growth of barley plants in relation to the presence of earthworms, microorganisms and the accumulation of vitamin B_{12} in experimental pots. Bacteria and actinomycetes are mainly responsible for the production of vitamin B_{12} in soil. The presence of earthworms, notably *Allolobophora rosea*, *A. caliginosa*, *Lumbricus rubellus* and *L. terrestris*, enhanced this microfloral activity. Two- to three-fold increases in numbers of microorganisms occurred in experimental pots, compared with controls (from which earthworms were excluded), after a period of 6–12 months. In the presence of earthworms, the amount of vitamin B_{12} in soil increased at least two-fold (and in some cases, between five- and seven-fold, depending on the number of earthworms used and the length of the experiment), and there were concomitant increases in the yield of barley. No significant differences could be detected between the various earthworm species and the accumulation of vitamin B_{12} in the soil. This is rather surprising in view of the fact that different 'life styles' were represented among the four species selected for this study. Thus, *L. rubellus* is épigé, *A. caliginosa* is endogé, while *L. terrestris* is anécique. It might be expected that, with different habitat preferences and different diets, these three ecological groupings would show different inter-relationships with soil microflora. However, Atlaviniyte and Daciulyte did not identify the microorganisms present, nor were they able to provide data correlating the numbers of microorganisms present at any one time with the level of vitamin B_{12}.

4 Blood and Breathing

4.1 Introduction

Earthworms have a 'closed' vascular system. Blood flows through vessels that are responsible for its distribution to all parts of the body. This system contrasts with that of molluscs and arthropods, for example, which is of the 'open' type. In these phyla, blood is pumped from a muscular heart to the tissues requiring food and oxygen and the removal of carbon dioxide and metabolic wastes. From thence it drains into a large blood space, the haemocoel, the main body cavity that surrounds the viscera. Subsequently, it returns to the heart through pores, termed ostia, in its walls (arthropods) or *via* a pulmonary vein (molluscs). This system is inefficient to the extent that there is a loss of pressure as blood drains into the haemocoele. Not so in earthworms where pressure can be maintained throughout the system.

4.2 Earthworm 'hearts'

The main structures responsible for creating vascular pressure are the so-called 'hearts' and the contractile dorsal longitudinal vessel. These parts of the vascular system are equipped with valves that allow the blood to flow only in one direction through them. The 'hearts' are situated in segments VII to XI, in members of the genus *Lumbricus* at least. They are contractile vessels that form dorso-ventral connections between longitudinal vessels running above and below the gut (Fig. 4–1). Their number varies in different groups of

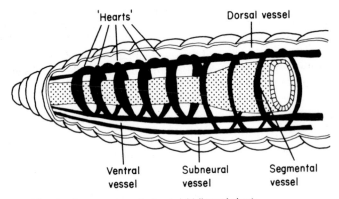

Fig. 4–1   The circulatory system of a lumbricid (lateral view).

earthworms; there are two to five pairs in megascolecids, for example (EDWARDS and LOFTY, 1977).

4.3 Vessels

It is not possible to distinguish 'arteries' and 'veins' in earthworm vascular systems. This is because there is no mechanism for the separation of blood rich in oxygen and nutrients from that which is poor in these commodities. Instead, there is a series of longitudinal vessels that distribute blood in various directions. The main vessels that do this in the earthworm body are the dorsal, ventral and subneural vessels (Fig. 4–1). The dorsal vessel is situated above the gut and is a single structure as a rule, although it is paired in some megascolecids and glossoscolecids. Blood flows forwards in this part of the system from posterior to anterior. The ventral vessel is located beneath the gut and conveys blood from the anterior part of the body to the posterior. The blood also flows from the anterior to posterior in the subneural vessel when this is present. Segmentally-arranged dorso-ventral vessels connect up these longitudinal vessels. In segments VII to XI these are the 'hearts' already mentioned, and they provide connections between the dorsal and ventral vessels. More posteriorly, the dorso-ventral connection is between the dorsal vessel and the subneural vessel, at least in *Lumbricus*. Other longitudinally-arranged vessels that complement those already mentioned in the distribution of blood to various parts of the body include the latero-neurals and the latero-oesophageals. The former are present in most earthworms (except the Megascolecidae). They are situated on either side of the ventral nerve cord and are closely associated with it along its entire length. The latero-oesophageal vessels run along the sides of the anterior part of the gut ending posteriorly about the level at which the dorsal/subneural connectives commence. The blood flows from anterior to posterior in both of these systems. From these main longitudinal vessels, and particularly from the ventral vessel, smaller branches arise that convey blood to the body wall.

This system, with other embellishments that have been detailed by EDWARDS and LOFTY (1977), provides for a transport system that can distribute food and oxygen to various parts of the body and remove waste products from these parts. Blood entering the typhlosole of the intestine from branches of the dorsal vessel is enriched with food material. This enriched blood then passes into the ventral vessel and is distributed to various parts of the body by paired ventro-parietal (or integumentary) vessels in each segment.

4.4 The respiratory system

In earthworms respiration is essentially by gaseous exchange across the body wall. This necessitates a richly vascularized epidermis to facilitate the exchange between the blood and the external environment. Such is provided, in *Lumbricus* spp. at least, by a network of looped capillaries that allow a flow of blood into and out of the body wall (Fig. 4–2). Control of this flow is mediated

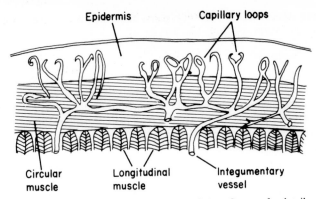

Fig. 4–2 Blood supply to the body wall of *Lumbricus*. See text for details.

by the vasoconstrictive action of adrenalin on the capillaries and dorsal vessel in *Glossoscolex giganteus* (JOHANSEN and MARTIN, 1966). These capillaries are the terminal portions of the ventro-parietal vessels that have their origins in the ventral vessel, as already noted.

In order for the general body surface to act as a respiratory organ it must be moist. The exchange of oxygen and carbon dioxide must occur in solution, since these gases diffuse through the body wall in a dissolved state. To facilitate this exchange, the body surface is kept moist by secretions from epidermal mucous glands, the dorsal pores and also the excretory discharge from nephridiopores.

This type of respiratory system relying, as it does, on passive diffusion of gases across the body wall is not the most efficient. It does not allow for the rapid penetration of oxygen from the external environment into tissues situated deep in the body, nor the removal of carbon dioxide from such tissues and its transport to the surface. However, the respiratory pigment haemoglobin is present.

4.4.1 Respiratory pigments

The blood of earthworms is red in colour due to the presence of the respiratory pigment haemoglobin in solution in the plasma. Without this pigment, the amount of oxygen dissolved in the body fluids would be directly related to the partial pressure of oxygen in the external environment. If this pressure is low, there may be insufficient oxygen in solution to satisfy the metabolic requirements. Haemoglobin is a protein that combines very readily with oxygen and acts as an agent in its transport and storage.

Haemoglobins occur in a wide range of animals, vertebrate and invertebrate. The efficiency with which they combine with oxygen varies considerably from one group to another. The haemoglobin of *Lumbricus terrestris* and *Allolobophora longa*, for example, shows a high affinity for oxygen in conditions approaching those encountered in Nature. This haemoglobin is fully oxygenated at pressures which would saturate only about 50% of human haemoglobin, for example (Fig. 4–3).

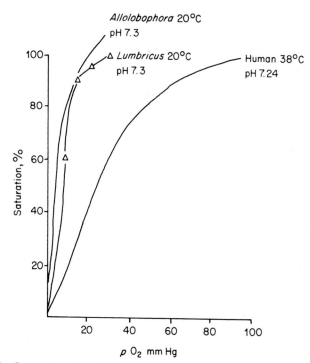

Fig. 4–3 Comparison of equilibrium curves for the respiratory pigments of *Allolobophora, Lumbricus* and human.

Earthworm haemoglobin combines preferentially with carbon monoxide under experimental conditions, even when oxygen is present in abundance. This combination blocks the normal functioning of haemoglobin as an oxygen acceptor, and permits the assessment of the importance of this pigment in normal respiration (GARDINER, 1972). In *Lumbricus terrestris*, for example, experimental blockage of haemoglobin in this way results in a significant drop in oxygen consumption at pressures above 8 mm Hg. This suggests that, under normal conditions, the transport of oxygen in combination with haemoglobin is a continuous process. Further, such experiments indicate that no more than half, at the most, of the oxygen utilized is carried in this way – the remainder is transported in solution in the coelomic fluid.

4.4.2 Carbon dioxide

The mechanisms for the removal of carbon dioxide from the earthworm body are poorly understood. This product of respiratory metabolism could, in theory, be transported in solution in the coelomic fluid to the tissues of the body wall, and thence to the exterior by passive diffusion down a concentration gradient. During this passage however, carbon dioxide combines with water molecules in the coelomic fluid to produce carbonic acid, a reaction that is catalyzed by the enzyme carbonic anhydrase. Carbonic acid dissociates to form hydrogen and bicarbonate ions, and the latter may combine with calcium ions to form insoluble calcium carbonate which passes out of the body through the gut with the faeces (see Chap. 3). Active calciferous glands (see p. 11) are important sources of the anhydrase enzyme and would seem to be involved in the removal of carbon dioxide from the body of those species that possess them. Little is known of the process in species that do not.

4.5 Ecological implications

The high affinity for oxygen of earthworm haemoglobin implies that oxygen acceptance occurs even when the external partial pressure is low. Such a condition may prevail in poorly ventilated soils. Earthworms are evidently adapted to colonize these through the special properties of their respiratory pigment. This is particularly true of deep-burrowing species (anécique) which rarely, if ever, venture to the surface. However, earthworms may experience conditions of intolerably low oxygen tensions, for example when the soil is flooded after rain. When this happens they tend to leave their burrows and emerge on to the surface to avail themselves of the higher oxygen levels in the atmosphere.

On the other hand, earthworms cannot turn to their advantage conditions of high external oxygen pressure since their haemoglobin is saturated quickly. In these circumstances, the oxygen that is available – over and above that which has been used to saturate the haemoglobin – can diffuse through the body wall and go into solution in the coelomic fluid. The transport of oxygen by this means is not a very efficient process, as we have already seen. The low respiratory rates that have been recorded for earthworms (SATCHELL, 1967) may reflect this fact. These rates can be an order of magnitude lower than those of other soil saprophages, for example enchytraeids and collembolans (WALLWORK, 1970). This fact must be taken into consideration when assessing the contribution made by earthworms to total soil community respiration. Estimates of this respiration provide an index of the extent to which various groups of soil fauna appropriate the energy available in the form of leaf litter and other particles of detritus. Expressed in these terms, the contribution made by earthworms is relatively low – only 8–10% of the energy input to an English woodland soil, according to SATCHELL (1967). The biomass of earthworms in such soils and those of calcareous grassland is high (Table 4–1).

Clearly, biomass alone is not a good index of the contribution made by these saprophages to total community metabolism.

Table 4–1   Biomass and respiration of three groups of soil fauna (from WALLWORK, 1970).

Group	Biomass (gm^{-2})	Respiration rate (μlO$_2$ mg^{-1} day^{-1})	
Lumbricidae	61	3.0	
Enchytraeidae	1.6	23.76	(max)
Collembola	0.2	24.0	(max)

5   Excretion and Water Relations

5.1   Introduction

The majority of annelids are aquatic. Polychaetes are almost exclusively marine, the leeches occur mainly in fresh water, and many of the oligochaetes live in fresh water. It may be concluded with a fair degree of certainty that annelids evolved in a watery environment, probably the sea, invaded fresh water and from here some enterprising groups moved on to land. Invasion of the land habitat, during the course of animal evolution, has proved to be a formidable task, to judge from the fact that only six phyla out of about thirty have attempted it with a good deal of success. These are the annelids, molluscs, crustaceans, uniramous arthropods, chelicerates and the amniote vertebrates. The problems involved in making this transition are common to all these groups, although they have been solved in different ways and with differing degrees of success. They are, basically: how to maintain body shape in a less dense medium; how to ensure that the developing embryo is surrounded by life-sustaining fluids; how to adapt respiratory systems to breathe air rather than water; how to cope with the wide range of dietary items that the terrestrial environment affords; how to achieve homeostasis of body fluids, a control of body salt and water content in an environment that is much less stable than that of aquatic habitats.

It is with this last problem that we are immediately concerned for it is, in many ways, the most crucial. To be successful on land, earthworms have to control their internal environment, and this means eliminating metabolic wastes without undue loss of water and essential ions. The conservation of water is a pre-requisite for living on land and since soluble salts may move in the same direction as water the problems of water conservation and ionic balance, although distinct, are closely associated. These functions are primarily the responsibility of the excretory system which, in the case of earthworms, has been inherited from their aquatic ancestors. This excretory equipment is not ideally suited to life on land, but earthworms have been able to adapt and improvise in this regard, as we will see.

5.2   Nephridia

The principal organs of excretion in earthworms are nephridia, literally from the Greek *nephros*, a small kidney. In its basic form, a nephridium is a simple tubular structure of ectodermal origin that conveys metabolic wastes from the coelomic fluid and blood to the exterior *via* an aperture, the nephridiopore. It is an organ that is paired and serially repeated down the length of the body in segmented worms.

This basic plan has been modified in a variety of ways in different groups of invertebrates, although all of these varieties can be referred to one or other of two major types: protonephridia and metanephridia.

Protonephridia occur, for the most part, in lower invertebrate phyla, such as the Platyhelminthes, Aschelminthes and Nemertea, although they are also present in molluscs and cephalochordates. They also occur, as do metanephridia, in polychaete annelids. Protonephridia are characterized by a simple tubular structure that terminates, internally, in solenocytes or flame bulbs. Solenocytes are embedded in mesoderm from which they obtain their excretory materials; these materials pass into the nephridial duct and thence to an excretory pore located in the body wall.

The variations on the protonephridial theme are numerous (Fig. 5–1).

(*1*)   Protonephridia can become intimately associated with the segmental genital ducts (coelomoducts) to form structures known as *protonephromixia* (Fig. 5–1a); examples of these are found in polychaetes.

(*2*)   In other coelomate animals, the nephridial duct may terminate internally in the coelomic space, rather than embedded in mesodermal tissue, and this type is termed a *metanephridium* (= metanephromixium; Fig. 5–1b).

(*3*)   The terminal portion of a metanephridium which is in contact with the coelomic fluid is a ciliated, funnel-like structure, termed a nephrostome. The development of the nephrostome varies in different groups of oligochaetes. In terrestrial forms the upper lip of this funnel is multicellular and horseshoe shaped (*meganephrostome*). In aquatic forms this upper lip is variously reduced to a few marginal cells (*mesonephrostome*) or is absent (*micronephrostome*).

(*4*)   The function of the nephrostome is to collect excretory materials from the coelomic fluid, and this funnel may be open or closed. If open, wastes from the coelomic fluid can pass freely into the nephridial tubule. If closed, these wastes must be filtered through the walls of the funnel into the tubule.

(*5*)   Nephrostomes are segmentally arranged, but within a particular segment these funnels may be single or multiple. Nephridia with multiple nephrostomes are termed *meronephridia*.

(*6*)   Basically, the nephridial duct opens directly to the exterior *via* a nephridiopore located in the body wall: the *exonephric* condition. Alternatively, this duct may discharge its contents into the alimentary canal: the *enteronephric* condition.

(*7*)   Finally, a condition occurs quite commonly in polychaetes in which the nephrostome disappears completely, the internal opening of the coelomoduct becoming completely fused with the internal end of the nephridial tubule. The urinogenital organ so formed represents a combination of ectodermal (nephridial) and mesodermal (coelomic) structures, and is termed a *mixonephridium* (= nephromixium; Fig. 5–1c).

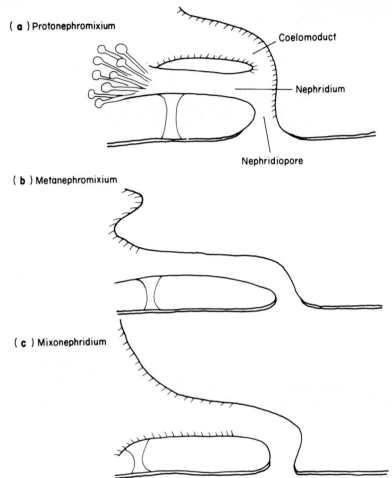

(**a**) Protonephromixium

Coelomoduct

Nephridium

Nephridiopore

(**b**) Metanephromixium

(**c**) Mixonephridium

Fig. 5–1   Variations on the protonephridial theme.

5.2.1   *Metanephridia*

In earthworms, the principal excretory organ is the metanephridium. Nitrogenous wastes are also eliminated through the body surface as muco-proteins, but the main avenue of nitrogen excretion is the nephridial system. Earthworm nephridia are completely divorced from coelomoducts. The latter maintain their integrity as gonadial ducts and are restricted to those few segments of the body in which the gonads are situated. In the *Lumbricus* nephridium, open nephrostomes are located in the segment immediately anterior to that containing the tubular portion of the metanephridium (Fig. 5–2). Beyond this, few generalizations can be made regarding other groups of earthworms.

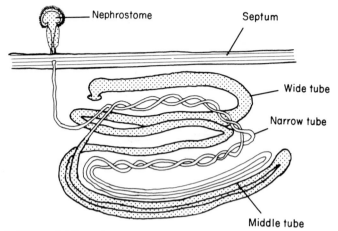

Fig. 5–2   The nephridium of *Lumbricus*.

For example, in the Indian earthworm *Eutyphoeus foveatus* there are both open and closed meronephridia within the same individual, whereas in members of the genus *Lampito*, also found in India, there are, characteristically, open meronephridia. A variant of the 'multiple', or meronephridial, type is the 'tufted' nephridium found in many members of the Megascolecidae and Glossoscolecidae (e.g. the genera *Pontoscolex* and *Pheretima*). Here the proximal part of the nephridium is profusely branched – a condition possibly resulting from the union of a number of meronephridia. All of these terminal branches discharge into a common segmental excretory duct that may be exonephric (*Pontoscolex*) or enteronephric (anterior and posterior segments of *Pheretima*). The terminations are open in *Pontoscolex*, closed in *Pheretima*.

5.2.2   Exonephridia and enteronephridia

The excretory system of earthworms belonging to the genus *Pheretima* is of special interest, for there are three functionally distinct groups of nephridia. An anterior group lies in segments IV, V and VI; the ducts from these run forward and empty their waste into the buccal cavity, to be voided through the mouth. These nephridia are often termed 'pharyngeal' to distinguish them from the posterior group commencing at the septum between segments XV and XVI. These empty their waste into a pair of longitudinal ducts that lie on the dorsal surface of the intestine. Segmental connections between these ducts and the lumen of the intestine provide pathways for waste materials to be passed into the gut and eliminated from the body through the anus. These two enteronephric systems are complemented by an exonephridial system consisting of numerous (200–250 per segment) integumentary nephridia located mainly in segments VII to XV.

A similar arrangement is found in members of the genus *Lampito*, except that here there is but a single longitudinal, median excretory duct that discharges, segmentally, into the intestine (Fig. 5–3). In addition, there are 20–24 pairs of independent nephridia in each segment, which extend along the septum and open by very small nephridiopores in the ventral body wall.

This theme, with minor variations, is repeated in other earthworms. Enteronephric systems are present in *Allolobophora antipae* and *Hoplochaetella khandalensis,* and here the posterior nephridia pass the waste materials into a pair of longitudinal canals that run along the inside of the body wall, emptying into the alimentary canal just in front of the anus (Fig. 5–4). *Megascolex cochinensis* also has an enteronephric system associated with meronephridia. The multiple nephridia drain into a pair of common ducts, one on the left and the other on the right, aligned at right angles to the long axis of the body. Each of these ducts curves dorsally, and the two unite above the intestine in the mid-line before entering, as a common duct, into the intestine (Fig. 5–5).

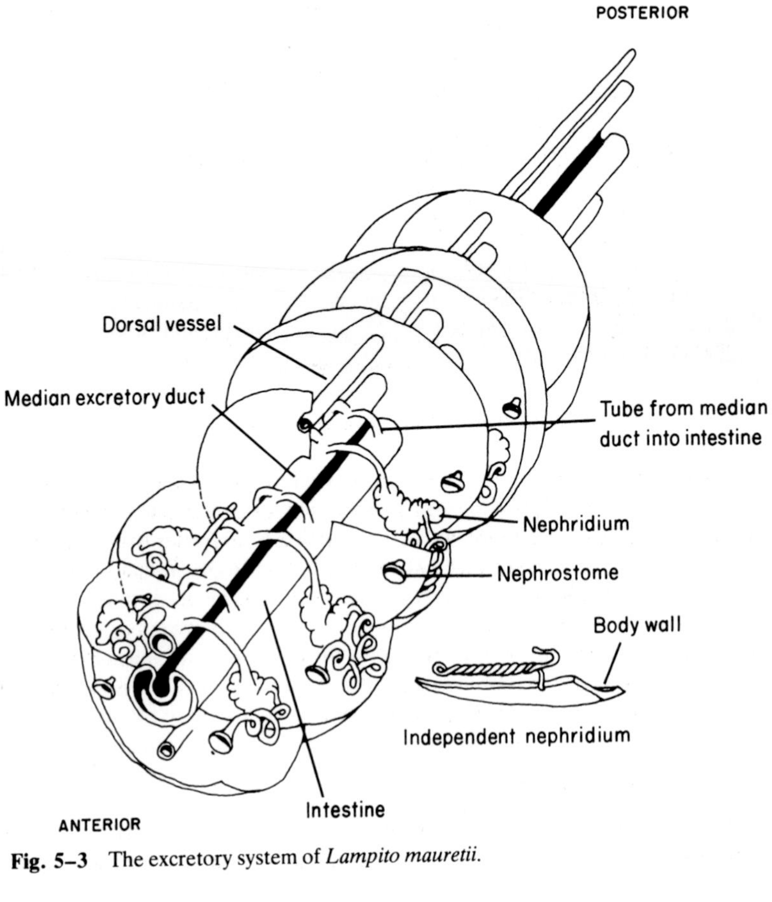

Fig. 5–3 The excretory system of *Lampito mauretii.*

ANTERIOR

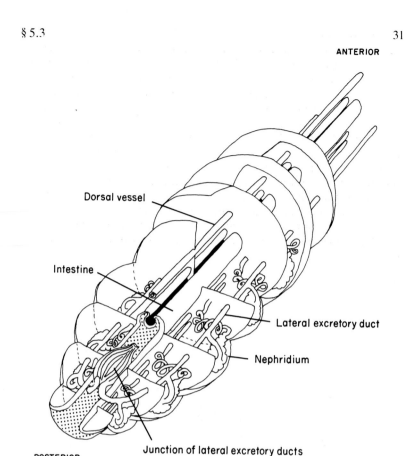

Dorsal vessel

Intestine

Lateral excretory duct

Nephridium

POSTERIOR

Junction of lateral excretory ducts
with opening into intestine

Fig. 5–4   The excretory system of *Allolobophora antipae*.

Earthworms of the genus *Lumbricus* have a simpler arrangement which is, perhaps, atypical. The metanephridium is exonephric and consists of a proximal, open meganephrostome, and the nephridial duct from this passes back through the septum and becomes structurally differentiated (see Fig. 5–2). The duct coils upon itself, becomes wider, and receives a rich supply of blood from the ventral vessel. The ventral vessel is the main avenue for oxygenated blood pumped backwards from the 'hearts'. This blood probably serves to promote metabolic activities associated with the uptake of water and salts from excretory wastes (see later, p. 35).

5.3   Excretory products

Basically, animal excretory products are nitrogenous in character and they can take one of three forms: (1) predominantly ammonia, an ammonium

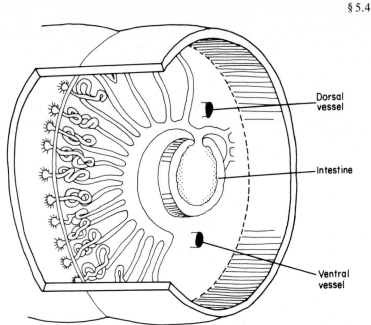

Fig. 5–5 The excretory system of *Megascolex*.

compound, or a derivative of such – this is *ammonotelic* excretion; (2) the main excretory product may be uric acid (*uricotelic* excretion); (3) urea may be the main excretory product (*ureotelic* excretion). These three categories are by no means absolute and, indeed, the urine of earthworms may contain ammonia, urea and, to a limited extent, uric acid although the relative proportions of these substances vary. They vary with the species of earthworm, with the season, the diet of the animal and with its nutritional state. Under normal conditions of adequate diet, members of the genera *Lumbricus*, *Allolobophora* and, less consistently, *Eisenia* excrete ammonia primarily, but switch to urea when starved (GARDINER, 1972). Comparisons between starved and fed individuals reveal that *Eisenia* excretes twice as much urea when starved, rather than fed, *Allolobophora* four times as much, and *Lumbricus* thirty to forty times as much. This change from ammonotelism to ureotelism is apparently controlled by enzymes of the urea cycle. These enzymes are evidently triggered into activity when the nutritional state of the animal falls, and are inhibited when this state rises. Since ammonia requires more water for its elimination from the body than does urea, the switch to ureotelism may be a device to restrict the loss of water from the body during periods of nutritional stress.

5.4 Origin of excretory products

Excretory products pass through the nephridial system either directly or by filtration through the nephrostome from the coelomic fluid, or filter through the walls of the nephridial duct from the blood system. These products therefore originate from two sources: coelomic fluid and blood. Specialized storage cells, derived from coelomic epithelium and closely associated with the wall of the intestine, have been implicated in this process. These constitute the chlorogogenous tissue that has been shown to accumulate varying amounts of ammonia and urea. However, it has still not been established whether these substances pass into the coelomic fluid by diffusion, or whether the chlorogogenous cells break loose, float freely in the coelomic fluid, and then disintegrate to release their excretory load.

5.5 Salts and osmoregulation

Earthworms maintain their internal osmotic pressure above that of their surroundings, a characteristic probably inherited from their freshwater ancestors. They excrete a hypotonic urine as a consequence of the resorption of salts through the nephridial wall. The concentrations of major inorganic ions in the blood of *Pheretima* have been estimated and compared with those of the polychaete *Arenicola* (Table 5–1).

The absolute and relative differences between the ionic composition of the blood of these two genera reflect the fact that *Arenicola* lives in a saline environment where sodium and magnesium ions are plentiful, whereas *Pheretima* behaves very much as a freshwater animal faced with the problem of a limited supply of these ions. Indeed, all earthworms surrounded, as they are, by a film of watery mucus can be said to live in a kind of freshwater environment.

Nevertheless, earthworms, unlike other more truly aquatic oligochaetes, have to face the problems of water loss. Even in the moist environment in which they live, earthworms are rarely completely hydrated. Members of the genera *Lumbricus* and *Eisenia* will increase their body weight by some 15% when taken from soil and placed in tap water for a period of 5 hours; on return to the soil they regain their original weight in a rather shorter period of time. This suggests that there is movement of water through the body wall. In the soil environment this movement is predominantly from the animal to the exterior. Other sites of water loss are the dorsal pores, the nephridia and the alimentary canal. These are also the avenues through which salts can be lost from the body. It has been demonstrated, for example, that earthworms can take up

Table 5–1 Ionic composition of the blood of *Pheretima* and *Arenicola*.

	mMl^{-1}			
	Na^+	K^+	Ca^{++}	Mg^{++}
Pheretima	43	7	5	7
Arenicola	459	10	10	52

chloride ions through the body wall against a concentration gradient when immersed in hypotonic salt solutions. Under natural conditions, water loss through the body wall would also be accompanied by losses of soluble salts. Detailed studies have also shown that in earthworms such as members of the genus *Lumbricus*, that have open nephridial systems, chloride, sodium and potassium ions enter the excretory tubule from two sources: from the coelomic fluid *via* the nephrostome, and from the blood by pressure filtration through the wall of the proximal narrow part of the tubule (Fig. 5–6).

How then do earthworms cope with these potential losses of water and salts? Several mechanisms are present that may help to solve these problems, and these mechanisms combine in different ways in different species. They can be divided, broadly, into two categories: behavioural and physiological. Behavioural adaptations are exemplified by the tendency of many species to select moist environments and to move deeper in the soil profile when the upper horizons become desiccated. Further, many earthworms, and members of the Lumbricidae are good examples, avoid acidic soils where salts are in short supply; their preference for calcareous grassland soils ensures that

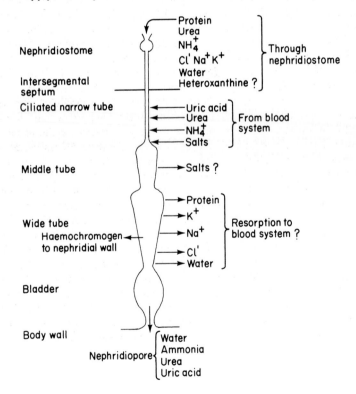

Fig. 5–6 Diagram summarizing possible mode of functioning of the oligochaete nephridia based on *Lumbricus* (after LAVERACK, 1963).

adequate amounts of calcium and potassium are present in their diet. The selection of moist environments, together with the watery envelope of mucus that surrounds the body, effectively reduces the water vapour gradient across the body wall, and also curtails the loss of salts in solution.

Loss of water through the nephridial system is unavoidable if the main excretory products are ammonia and/or urea, since these soluble products require water for their elimination from the body. However, earthworms have mechanisms for minimizing this source of water loss. The ability to excrete nitrogenous waste in the form of uric acid presumably would stand them in good stead during times of extreme water stress, although there is no strong evidence that they possess this ability to any marked degree. But there are other physiological adaptations that have been documented. In *Lumbricus*, as already noted, the nephridial tube is differentiated into regions of differing diameters; these regions are also functionally differentiated. Freezing point determinations have shown that the composition of the urine changes markedly during its passage from the nephrostome to the bladder (RAMSAY, 1949). In the nephrostome region, the urine is iso-osmotic with the coelomic fluid, but by the time it reaches the bladder it is hypo-osmotic not only to the coelomic fluid, but also to the blood. The major site of resorption of water, protein, calcium, sodium and potassium is the richly vascularized wide portion of the nephridial tube (Fig. 5–6). Here, about 80% of the solutes in the urine are returned to the blood against concentration gradients.

A further refinement is the development of enteronephric systems. The discharge of urine into the gut, rather than directly to the exterior through the body wall, allows for further resorption of water and salts through the gut wall. These conservation measures are particularly important for tropical earthworms that have to overcome the problems of desiccation during the dry season. The enteronephric system of members of the genus *Pheretima*, already described, allows these earthworms to lose only minimal amounts of water through the mouth and body wall, and to void faeces that are practically dry when they are living in dry sandy soil. During the rainy season, when the worms face the danger of over-hydration, water can be removed from the body not only through the enteronephridia but also through the numerous integumentary exonephridia. As mentioned earlier, earthworm nephridia do not associate with coelomoducts, as they do in many polychaetes. This separation of excretory and reproductive functions would seem to be a pre-requisite for the development of an enteronephric system and the flexibility that this conveys – there would appear to be no virtue, for example, for having a mechanism for shedding gametes into the gut. This flexibility allows earthworms, such as *Pheretima*, to control to a large degree their internal osmotic environment in the face of external environmental stresses.

6 Sense and Sensitivity

6.1 Introduction

The development of bilateral symmetry, first encountered in the Platyhelminthes and elaborated in the Annelida, is a significant advance over the radial arrangement found in the Coelenterata. Among other things, it opens up the possiblity of concentrating structures designed to collect information from the environment where they will do most good, i.e. at the anterior end of the body. This process, known as cephalization, is very much a feature of the annelid/arthropod line of evolution. It culminates, in the arthropods, in the localization of organs of special sense, eyes and antennae in particular, in a clearly demarcated region: the head. Annelids do not have a definite head region but cephalization is clearly expressed by the concentration of sensory structures, receptors, at the anterior end of the body.

6.2 Receptors

Earthworms are sensitive to chemicals in their environment, to touch and to light. Much of the life style of these animals, their habitat selection, dietary preferences and activity patterns, can be explained in terms of this sensitivity.

Chemoreceptors occur all over the body of earthworms as epidermal tubercles and as free nerve endings. Each tubercle is a cluster of sense cells with nerve connections and with extensions in the form of sensory hairs that protrude through the cuticle (Fig. 6–1). Structurally these receptors are similar to the chemosensors of planarians, on the one hand, and the taste buds of vertebrates, on the other. Despite the simplicity of their design, Natural Selection has not been able to improve on it. As might be expected, these receptors are concentrated on the prostomium, the wall of the buccal cavity and the pharynx. Here their main function is gustatory; they taste the environment and select palatable food items for ingestion. As noted in Chapter 3, British earthworms, at least, feed selectively on leaf litter rich in nitrogen and sugars. In the case of *Lumbricus terrestris*, sensitivity to sugar is apparently lodged in the prostomium (see GARDINER, 1972) where as many as 700 taste receptors may be present in a square millimetre. Sensitivity to salt concentrations is much more widely distributed over the entire body surface however, and this is probably a function of free nerve endings in the epidermis.

The ability to respond to soil acidity is variously developed in different species and, again, this phenomenon has been most extensively studied in British lumbricids. The evidence for this response is both circumstantial and direct. The distribution patterns of *Dendrobaena rubida* and *D. octaedra*, for

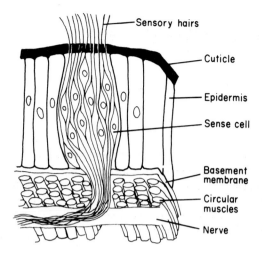

Fig. 6–1 Epidermal chemoreceptor in *Lumbricus.*

instance, indicate that these species can tolerate acid conditions. By the same token, *Allolobophora longa, A. caliginosa* and *A. chlorotica* are acid-intolerant (SATCHELL, 1955). More directly, LAVERACK (1961) tested the responses of various species to a range of buffer solutions. He found that *Allolobophora longa, Lumbricus terrestris* and *L. rubellus* exemplified a gradient of sensitivity ranging from the acid-intolerant *A. longa* to the tolerant *L. rubellus*, with *L. terrestris* occupying an intermediate position.

The detection of soil acidity is evidently a function of free nerve endings distributed over the general body surface. LAVERACK (1961) discovered that this was, indeed, a direct response to pH and not to calcium levels in the soil. Soil pH and calcium level are related phenomena but they are not synonymous. Environmental calcium is evidently important to many earthworms, possibly because of its role in carbon dioxide removal (see Chap. 4). However, there is no evidence that worms respond directly to calcium levels in the soil the way that molluscs do, for example, or woodlice and millipedes. Maybe they are guided to calcium-rich environments by their sensitivity to pH. This sensitivity is the property of receptors and segmental nerves that are, apparently specific for this purpose. They are distinct, for example, from receptors that detect levels of sodium chloride, quinones and sugars.

There is still much to be learned about chemosensitivity in earthworms. We know, for example, that certain species feed on cattle dung and are attracted to it when it is fresh (BOYD, 1958). Such species include the acid-tolerant *Dendrobaena octaedra, Lumbricus castaneus* and *L. rubellus*. It is unlikely, however, that this is a simple and direct response to low pH, since the acid-intolerant *Allolobophora chlorotica* also reacts in this way while the acid-tolerant *Dendrobaena rubida* does not. In any event, cattle dung is not particularly acidic and its attractive properties may reside in its nitrogen content, for example, or some other property of its biochemistry. This remains to be determined. Again, *Eisenia foetida* is habitually associated with compost, *E. lucens* with decaying wood. Both of these species are probably responding to the biochemistry of these particular environments, but exactly how remains uncertain.

Tactile receptors are of particular importance to animals that burrow; they need to be able to sense the burrow around them if they are to orientate properly. Tactile sense is also needed if the worm is to re-locate its position within a burrow after a foray on the surface, although many burrowing species rarely emerge completely. Such receptors are distributed over the general body surface, and the information they obtain is relayed to the central nervous system by segmental nerves. In all probability, this tactile sensitivity is a function of free nerve endings.

Photosensitivity in earthworms is of the diffuse kind. Photoreceptors in the form of discrete eyes do not occur, although lens-like structures in the epidermis and dermis are found at the anterior end, in particular on the prostomium. The absence of true eyes is undoubtedly associated with the nocturnal habit, and *Pheretima* spp. are completely photonegative. However, this response is by no means universal since *Lumbricus terrestris* is photopositive to very weak light sources (EDWARDS and LOFTY, 1977). This response may allow these worms to emerge from their burrows on moonlit nights. The control and co-ordination of the light response have been reviewed by EDWARDS and LOFTY (1977) who conclude that the central nervous system is involved.

6.3   Central nervous system

The nervous system of earthworms is centralized, i.e. neurones and their extensions are concentrated into cords that run parallel to the long axis of the body. At intervals along these cords are ganglia; discrete collections of nerve cells contained within a connective tissue sheath. The cell bodies of these neurones are located around the periphery of the ganglionic mass, while the central portion consists of dendrites and the terminal branches of axons. This central portion is also the site of synaptic connections (see below).

Two developments have been significant in influencing the character of the annelid central nervous system. *First*, cephalization has, in some cases, produced a concentration of ganglionic masses anteriorly, resulting in the formation of a 'brain' (cerebral ganglion) which is situated dorsal to the gut. In *Lumbricus* spp. this 'brain' is located in the third body segment (Fig. 6–2) and represents the coalescence of nerve elements from the first three body segments. This condensation may not be apparent in other genera, however, where the cerebral ganglion is situated in the first body segement. Whatever its situation, the cerebral ganglion connects with the rest of the central nervous system, which runs ventral to the gut, *via* nerve tracts termed connectives or commissures. These connectives circle the oesophagus and may, themselves, be ganglionated. *Second,* metameric segmentation has produced a serial

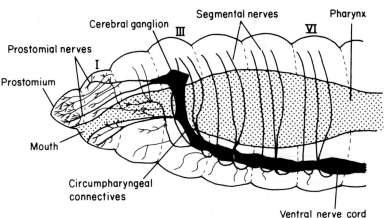

Fig. 6–2   Anterior part of central nervous system of *Lumbricus* (lateral aspect).

repetition of tissues and organs that is nowhere better expressed than in the central nervous system of annelids. The serial repetition of nervous tissue expresses itself in a chain of segmental ganglia, each ganglion linked by connectives, to form a ventral nerve cord. In the primitive condition, the ventral nerve cord is double. However, the two cords are fused together in earthworms to form a single chain. Metamerism is also reflected in the segmental arrangement of nerves issuing from the nerve cord. The number of these in each segment varies with the species. In *Lumbricus terrestris* for example, there is an anterior, median and posterior nerve on each side of the mid-line in each segment, and each of these nerves has a dorsal and ventral branch.

A feature of the nervous system of annelids, which they share with the molluscs and arthropods, is the presence of a stomodael system. This is a peripheral system of plexuses innervating the anterior part of the alimentary canal and connected to the cerebral ganglion and the anterior ganglia of the ventral nerve cord. In all probability sensory, motor and association neurones are present in this system. The function of this plexus is not fully understood, but it may well be involved in the maintenance of muscle tone and the secretion of digestive enzymes (GARDINER, 1972). It is well developed in oligochaetes.

Quite independent of the central nervous system in *Lumbricus* is a sub-epidermal nerve plexus that contains both sensory and motor neurones. This plexus innervates the circular muscles of the body wall, with which it combines to produce local responses outside the control of the central nervous system.

Giant nerve fibres, each consisting of a number of axons closely pressed together, run along the ventral nerve cord from the pharyngeal region to the posterior segments (Fig. 6–3). These axons are not continuous; they have septal synapses along their lengths. However, they are still able to transmit motor impulses at great speed, since these impulses by-pass the association

Lateral giant fibre   Median giant fibre

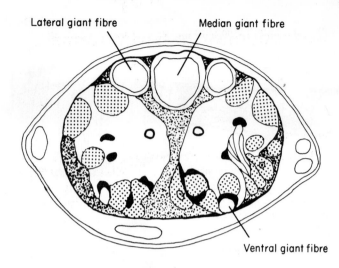

Ventral giant fibre

Fig. 6–3   Transverse section of ventral nerve cord.

neurones occurring more generally in the cord. The most conspicuous giant fibres are the large median and two smaller laterals located in the dorsal part of the ventral nerve cord. A pair of small ventrals in some species of *Lumbricus* has also been reported (GARDINER, 1972; EDWARDS and LOFTY, 1977). It seems likely that the large median fibre allows motor impulses to be transmitted only from anterior to posterior, while the smaller laterals transmit in the opposite direction. There are some conflicting reports in the literature concerning the speed with which these fibres transmit impulses. MEGLITSCH (1967) quotes a figure of up to 45 m sec^{-1}, while GARDINER (1972) states that impulses pass along the median fibre at speeds of between 17 and 25 sec^{-1}, and rather more slowly (7 to 12 m sec^{-1}) in the lateral ones. EDWARDS and LOFTY (1977) put the speed at the clearly impossible figure of 600 m sec^{-1} and do not distinguish between the median and lateral fibres in this regard.

Despite this confusion, it is generally agreed that the giant nerve fibres play an important part in relaying information rapidly along the earthworm's body. They are responsible for the rapid withdrawal of the animal into its burrow when subjected to unfavourable stimuli. This reaction is pronounced in anécique species in which the musculature for longitudinal contraction of the body is well developed. It may be less important for épigés and endogés, for this musculature is only weakly developed in these (BOUCHÉ, 1977). These considerations may go some way towards explaining the conflicting statements concerning the speed at which giant fibres transmit impulses.

7 Reproduction in Earthworms

7.1 Introduction

Many earthworms reproduce sexually, others are parthenogenetic to varying degrees. Sexual forms are hermaphrodite, i.e. male and female systems occur in the same individual. This is a departure from the typical polychaete condition in which the sexes are separate. Further, oligochaetes are not viviparous whereas some polychaetes are.

7.2 Reproductive systems

The serially-repeated arrangement of gonads that occurs in certain polychaetes has undergone substantial modification in terrestrial oligochaetes. Gamete-producing cells (gametocytes) that are derived from coelomic epithelium are concentrated and localized, as testes and ovaries, in a few body segments. In *Lumbricus*, for example, there are two pairs of testes on the posterior faces of the septa separating segments IX and X, and X and XI. The ovaries are located on the posterior face of the septum separating segments XII and XIII. Gametes (sperm) produced by the testes are not liberated freely into the coelom, since these testes are contained within sacs that are extensions of the septa IX/X and X/XI. These testes sacs are associated with, and provide continuity between, three pairs of pouch-like expansions of the body wall, termed seminal vesicles, located in segments IX, XI and XII. Sperm accumulate and mature in these pouches before being conveyed along a pair of vasa efferentia, anterior and posterior, that shortly unite as a common duct, on each side of the mid-line, the vas deferens. Sperm accumulated in the anterior and middle seminal vesicles are collected by the anterior vas efferens, while the posterior vas efferens collects from the posterior seminal vesicle. After entry into the common vas deferens the sperm are subsequently discharged to the exterior via male openings (gonopores) located on the ventral side of segment XV. In many earthworms (but not lumbricids) the distal portion of the vas deferens is equipped with prostate glands which produce fluid that facilitates the transfer of sperm. The male openings in lumbricids are readily visible to the naked eye as they are surrounded by swollen lips.

Eggs liberated from the ovaries are swept into the oviducts through ciliated funnels that open directly into the coelom. These oviducts pass through the septum into segment XIV and open on the ventral surface of this segment as inconspicuous pores. This arrangement of reproductive structures, typical of members of the genus *Lumbricus*, is depicted in Fig. 7–1.

Vasa deferentia and oviducts are coelomoducts and are derived from

(a)

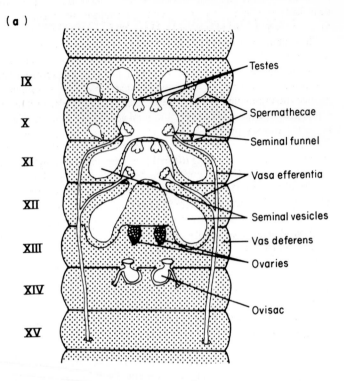

IX

X

XI

XII

XIII

XIV

XV

Testes

Spermathecae

Seminal funnel

Vasa efferentia

Seminal vesicles

Vas deferens

Ovaries

Ovisac

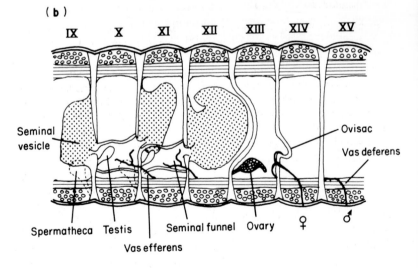

(b)

IX   X   XI   XII   XIII   XIV   XV

Seminal
vesicle

Ovisac

Vas deferens

Spermatheca   Testis   Seminal funnel   Ovary   ♀   ♂

Vas efferens

Fig. 7–1 Reproductive system of *Lumbricus*. (a) Dorsal; (b) lateral.

mesoderm. They are not associated with nephridia as they are in many polychaetes where the urinogenital structure so formed is termed a mixonephridium (see Chap. 5, and Fig. 5–1c).

Hermaphrodite animals, such as earthworms, run a genetic risk: that of in-breeding. In the short term, this is not serious, since in-breeding tends to conserve those characteristics that have survived the severe scrutiny of the Natural Selection process. In the long term, however, hermaphroditism can reduce the opportunities for genetic change, in particular for the translocation of genetic information from one parental genotype to another. In other words, the evolutionary potential of the group as a whole is seriously impaired if self-fertilization becomes the rule. Earthworms have solved this problem, for the most part, by separating the male gametes, produced by one partner, from the eggs produced in this same individual. In many oligochaetes this has been accomplished by the development of spermathecae, or seminal receptacles. These take the form of spherical pouches, lying in the coelom and opening directly to the exterior. There are two pairs of these structures in *Lumbricus,* situated in segments IX and X. Their function, as the name suggests, is to receive sperm from the partner during copulation.

This arrangement of reproductive structures is typical of lumbricids and glossoscolecids although the number and location of spermathecae are variable in these families and, indeed, these receptacles may be lacking in some cases (see below). In addition, the male pore is frequently situated far to the posterior of the body in the Glossoscolecidae. Other variations on this theme are shown in Fig. 7–2.

7.3 Copulation and cocoons

Earthworms that reproduce sexually usually cross-fertilize. Pairing occurs at night as a rule, when two worms position themselves head to tail with ventral surfaces adposed. This usually brings the openings of the vasa deferentia of one worm opposite to the openings of the spermathecae of its partner (lumbricids are exceptions). Sperm is then discharged into these seminal receptacles. In the Lumbricidae the discharged sperm travels along a pair of longitudinal grooves on the ventral surface to the spermathecal openings of the recipient. During this exchange the worms adhere closely to each other – the chaetae of one are embedded in the body wall of the other. This close contact is re-inforced by the secretion of mucilaginous girdles around each worm that originate from the region of the clitellum. The clitellum is a glandular development of the epidermis that varies in its extent from one group of earthworms to another. In the Lumbricidae particularly, the position of the clitellum is relatively constant within a species, and this can be used for identification purposes (see GERARD, 1964). Frequently associated with the clitellar region are glandular thickenings, situated ventrally – the tubercula pubertatis. These thickenings take various forms: longitudinal or transverse ridges extending over several segments; pads or tubercles that may be segmentally arranged. Their function is not certain,

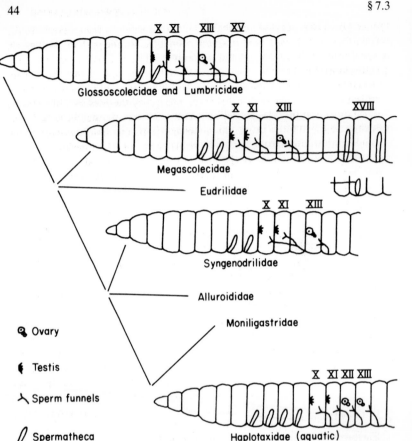

Fig. 7–2 Variations in the arrangement of reproductive structures in oligochaetes.

although they may produce secretions that assist in keeping worms together during copulation.

One variant of this copulatory theme perhaps deserves a mention. *Eutyphoeus waltoni* is an octochaetid occurring in India that apparently practises internal insemination. The swollen lips of the male pores are developed as an intromittent organ that can be inserted directly into the opposing spermathecal openings (BAHL, 1927). The eudrilid *Schubotziella dunguensis* apparently possesses a similar device for sperm transfer (EDWARDS and LOFTY, 1977).

Once the exchange of sperm has been completed, the two worms separate. Each then embarks on the production of a cocoon in which the eggs are laid. This cocoon is basically a mucous bag that is secreted by the anterior segments and the clitellum. The mucous envelope is passed forwards by reversed peristaltic movements of the body wall musculature. As it passes over the female gonopores, eggs are discharged into it. Further to the anterior are the spermathecae with their store of sperm from the erstwhile partner. This sperm is shed into the cocoon and the eggs are fertilized. Fertilization is, then, essentially external and successive cocoons are produced until the store of eggs and sperm has been exhausted. Some eudrilids apparently practise internal fertilization (SIMMS, 1964). This is accomplished by the presence of a special 'fertilization chamber' that receives sperm from the spermathecae and eggs from the ovisac. This is an exception, however, and the majority of terrestrial oligochaetes produce cocoons, and these are shed from the anterior end of the worm. As the cocoon passes over the prostomium its ends seal up to produce a tight envelope. The wall of this envelope subsequently hardens as it dries out.

7.4 Development after fertilization

Usually only a fraction of the eggs present in a cocoon will develop further. This may be because not all of the eggs are fertilized although there may be other causes of egg mortality, as yet unknown. The eggs that do survive – and, in British species at least, there is frequently only one – are nourished by nutritive albumen secreted by the glandular tissue that also secretes the cocoon. Further development of the egg is by spiral cleavage.

Young earthworms hatch from cocoons at times varying from several weeks to several months in Britain and North America (VAIL, 1974). There is some dispute over whether these juveniles emerge with the full complement of body segments, or whether they add segments as they grow to maturity. It would appear that species differ in this respect. Certainly, worms that have lost segments can, in some cases, regenerate all or most of the missing ones. This ability is much better developed in anécique species than in épigés and endogés (BOUCHÉ, 1977; see also Chap. 8).

The onset of sexual maturity is signalled by the appearance of the clitellum – worms without this cannot reproduce. Subsequently, at the end of the reproductive life, worms become senescent and the clitellum regresses and may disappear completely.

7.5 Parthenogenesis

Not infrequently part or all of the male reproductive system is lacking or infertile, and egg development occurs without fertilization. Parthenogenesis is then obligatory. In Britian, *Allolobophora rosea*, *Dendrobaena rubida tenuis*, *Octolasion* spp. and *Eiseniella* spp. fall into this category, as do *Aporrectodea trapezoides* and *Bimastos parvus* in North America (SATCHELL, 1967; JAENIKE and SELANDER, 1979). Indeed, parthenogenesis is widespread in North American lumbricids, with 17 out of 33 species being mainly or obligatory parthenogens (JAENIKE and SELANDER, 1979). Exceptions to this both in Britain and North America include most of the species belonging to the genera *Lumbricus* and *Allolobophora*; sexual reproduction is the rule here. The occurrence of parthenogenetic polymorphism has also been described in North

American lumbricids (JAENIKE and SELANDER, 1979). *Dendrobaena octaedra, Dendrodrilus rubidus* and *Eiseniella tetraedra* show this; some forms (or morphs) have well developed male organs, whereas other morphs of the same species have reduced testes, seminal vesicles and spermathecae.

7.6   Seasonality and aestivation

Reproduction and cocoon production can occur throughout the year, although in such British lumbricids as *Allolobophora caliginosa* and *A. chlorotica* these activities reach a peak in late spring and early summer (GERARD, 1967). During dry periods many lumbricids aestivate. No cocoon production occurs during this period of quiescence. *Allolobophora longa* and *Allolobophora nocturna* probably undergo obligate aestivation, while *Allolobophora caliginosa, A. chlorotica* and members of the genera *Eisenia* and *Octolasion* may show facultative quiescence. Cocoon production is limited to the rainy season in parts of the world where there are sharply defined wet and dry seasons. In India for example, maximum cocoon production by lumbricids inhabiting pasture soils occurs in late October and early November (DASH and SENAPATI, 1982). In the semi-arid steppe grassland of U.S.S.R. *Eisenia nordenskioldi* restricts its reproductive activity to the period of the summer rains: mid-July to the end of August (POKARZHEVSKIJ and TITISHEVA, 1982). Similar patterns of seasonal activity are shown by African eudrilids (MADGE, 1969).

8 Earthworm Life Styles

8.1 Introduction

The idea that different groups of earthworms have different life styles was introduced in Chapter 2 in relation to burrowing activity. We can now return to this theme and expand on it since it provides a convenient framework for an overall synthesis of earthworm biology.

8.2 *r* and *K* selection

It has become a common practice among ecologists in recent years to identify the life style of an animal as the product of one of two kinds of selection, termed *r* and *K* (PIANKA, 1970). *r* selection operates in patchy, heterogeneous, and unpredictable environments where populations undergo erratic changes in size in an opportunistic manner. Reproductive rates are high to counteract high juvenile mortality, and physical factors, particularly those associated with climate, are the ones that determine the life style. Other consequences that follow from this type of selection are: small individual body size, large numbers of offspring and rapid development to maturity. *K* selection occurs where physical effects are predictable and benign – where climatic changes have little influence on population size or life style. Such populations are usually crowded, competition is intense and reproductive rates are kept low by this competition which increases with density (i.e. density-dependent population regulation). Development to maturity is long as individual body size tends to be large.

8.3 Environmental predictability

Patches of leaf litter lying on the soil surface, dung pats and decaying logs provide heterogeneous environments for earthworms. They are subject to much greater fluctuations in temperature and moisture than the deeper soil layers. These are precisely the conditions in which *r* selection would operate, and we can expect that species classified, ecologically, as épigé or endogé would exhibit many of the features associated with this type of selection. Deep-burrowing anécique species, on the other hand, could be expected to show the characteristics of *K* selection. They can avoid the stresses that the unpredictable environment at or near the soil surface imposes. Since burrowing ability varies among earthworms, this criterion can be used as a first approximation in defining those that are likely to be products of *r* selection, and those that are *K* selected.

8.4   Earthworms as 'pioneers'

It will be appreciated that *r* selected species will frequently be early colonizers in newly-created environments and we can now examine the extent to which earthworms qualify for this role.

Evidence that earthworms colonize 'fresh' soils is provided by studies on Dutch polders (van RHEE, 1969). These are soils that have been reclaimed from the sea and, initially at least, are highly saline. Apparently this does not present an ecological barrier to the establishment of earthworms. Indeed, various species have been introduced deliberately into these soils in an attempt to improve their texture and render them suitable for agriculture. Successful inoculations have been carried out with *Allolobophora caliginosa* and *A. chlorotica*. Moreover, natural immigration by *Lumbricus rubellus* from adjacent 'old' soil has resulted in this species becoming numerically dominant in polder soils. All three of these species are considered to be either endogé or épigé and their life style, certainly as far as their colonizing ability is concerned, is consistent with *r* selection.

The same is true of species such as *Lumbricus festivus, L. castaneus, Dendrobaena octaedra, D. rubida* and *Bimastos eiseni* which, along with *L. rubellus*, are early colonists of cattle droppings (SVENDSEN, 1957a, b). These species have a relatively high dispersal capability that is a function, at least in part, of their high respiration rates (BOUCHÉ, 1977). This is another characteristic feature of *r* selection.

K selected species are usually considered to be specialists. They rarely leave the safety of their burrows completely except, possibly, to copulate (*Lumbricus terrestris*), or in the event of flooding. The fact that the haemoglobin of *L. terrestris* becomes rapidly saturated with oxygen at low tensions (see Chap. 4) explains why such species are relatively lethargic. Such features do not suggest the kind of mobility that would allow for rapid dispersal.

8.5   Population processes

For several decades, ecologists have been pre-occupied with the factors that determine the size of animal populations. Controversy has been generated and this frequently polarizes around two different, and apparently opposing, schools of thought. Proponents of the 'climatic' school, following ANDREWARTHA and BIRCH (1954), maintain that population levels are the outcome of physical (climatic) effects on mortality and natality. The operation of these effects will cause populations to increase during times of plenty, and decrease during times of stress. The alternative view is that as a population increases in size it encounters resistance to further expansion from biotic factors in the environment, such as competition and predation. This resistance intensifies as the size of the population increases and relaxes as population size declines. In other words, populations are *regulated* within rather narrow, circumscribed size limits. This size approximates to that which can be supported by the environmental resources (e.g. food, shelter, nesting sites, mates) – the *carrying capacity* of the environment.

The ideas embodied in *r* and *K* selection suggest that these two alternative explanations are not mutually exclusive, and earthworms may provide a good illustration of this. Simply stated, species that are considered as épigé or endogé are likely to be *r* selected as far as their population dynamics are concerned, giving support to the 'climatic' school. Anéciques, as a product of *K* selection, can be expected to experience density-dependent constraints on population sizes. What evidence is there for this?

8.5.1   Earthworms as pests

In the first instance, there is very little information on long-term population changes in individual species of earthworms. *r* selection is associated with a 'boom-or-bust' phenomenon which is expressed in erratic fluctuations in population density that may lead to the species in question achieving 'pest' status. *K* selection, on the other hand, implies a more measured response, in terms of population density, to environmental pressures. In the absence of concrete data, we must resort to circumstantial or anecdotal evidence to identify these strategies.

According to EDWARDS and LOFTY (1977), the species that can reach pest proportions in arable and grassland soils are the shallow-burrowing épigés and endogés. This is consistent with the idea that these groupings are *r* selected. Further, épigés and endogés, such as *Lumbricus castaneus*, *L. rubellus* and *Dendrobaena subrubicunda*, have greater potential than anéciques for population explosions, since they produce many more cocoons each year (Table 8–1).

Table 8–1   Number of cocoons produced in a year by various species of British lumbricids, based on the output from one individual in each case. (After EVANS and GUILD. 1948.)

Species	Number of cocoons
Épigé	
Lumbricus rubellus	79
L. castaneus	65
Dendrobaena subrubicunda	42
Endogé	
Allolobophora caliginosa	27
A. chlorotica	25
Anécique	
Octolasion cyaneum	13
Allolobophora nocturna	3
A. longa	8

8.5.2   Avoidance and tolerance

Cocoon production is the principal way in which these shallow-burrowers survive during periods of environmental stress. *Lumbricus castaneus* is a case in point; it survives through the summer drought in western Europe solely as eggs protected in cocoons. This is indicative of the fact that its life history tactic has, through the process of Natural Selection, become geared to respond to periods of climatic stress by avoiding them as active stages. This is consistent with *r* selection. Typical anéciques, such as *Lumbricus terrestris* and *Allolobophora longa*, pass through periods of environmental stress in a state of inactivity or, at the most, diapause. This requires a much less drastic adjustment of the life history pattern and as a consequence, they can be considered to be more tolerant of environmental changes. In other words, anéciques do not have to 'track' their environment in the way that épigés do. This is an attribute of *K* selection.

8.5.3   Competition

Deep-burrowing species, such as *Allolobophora longa, Lumbricus terrestris*, megascolecids and microchaetids, arguably are not exposed to climatic stress for most of their lives. They can live deep in the soil where they are shielded from the often violent fluctuations in temperature and moisture that occur at the surface. On the other hand, food material in the form of decaying leaf litter may be limiting at the depths to which these anéciques burrow. This may evoke some measure of competition between these subterranean species, although little attention has been paid to this topic so far. Certainly, experience with subterranean mammals indicates that competition can be intense in this kind of environment (NEVO, 1979). The fact that deep-burrowers, such as *Lumbricus terrestris, Allolobophora longa* and *Octolasion cyaneum*, emerge from their burrows, even if only partly, to feed, indicates that food resources incorporated in the soil profile are inadequate to satisfy the worms' requirements. This is circumstantial evidence that anéciques are subjected to competitive interactions, and this is a consequence of *K* selection.

8.5.4   Territorial behaviour

Territorial behaviour is also a feature of individuals and species that are potential competitors. The creation of territories may allow for the equitable allocation of resources, thereby reducing actual competition. It may be stretching the definition of a 'territory' too far to apply it to the semi-permanent burrows of anéciques (since these are not defended), but these refuges do allow for 'spacing out' that may minimize inter- and intra-specific encounters.

8.5.5   Predation

Predation often acts in a density-dependent manner to regulate population size of many herbivores and saprophages. Since earthworms fall into the latter category, we may usefully enquire whether this is a factor with which they have to contend. The answer to this question is, undoubtedly, yes, but how they cope with it varies from one ecological group to another. The inability of surface-dwelling species to burrow places them at the mercy of predatory birds, and predation is likely to be intense at certain times of the year. Selection has favoured small body size and a coloration that blends with the background in these épigés; these features can be considered as products of *r* selection. Anéciques, on the other hand, have a relatively large body size. They are subject to predation by small mammals in their subterranean burrows and by birds when they emerge on to the soil surface. Their defence against this type of predation is mainly two-fold. Firstly, the ability to retreat rapidly into the burrow – this, by virtue of well-developed longitudinal muscles in the body wall; second, the capacity to regenerate amputated segments – a capacity not present or only weakly developed in most épigés. These features indicate selection for a response to predation that may minimize its effect on population size without eliminating it completely as a method of regulation. Such a measured response to a potential mortality factor indicates a *K* selected system.

8.5.6   *Efficiency* versus *production*

The large body size of most anéciques implies a relatively long period of development to maturity. Although this period varies with temperature, it can be up to 2 yr in *Allolobophora rosea* (PHILLIPSON and BOLTON, 1977). As a consequence, such forms will channel most of the energy acquired in their food into growth of body tissue – relatively little will be diverted into reproduction. In other words, the accent is on efficient utilization of energy by individual animals, rather than the production of offspring that may or may not survive. It follows from this that anéciques will produce few offspring, but these will have high probabilities of survival. These are all features of *K* selection. This is confirmed by the data presented in Table 8–1: namely that anéciques produce far fewer cocoons annually than do épigés. The probability of survival of the young hatching from these cocoons is not known. It may be inferred, however, that it would be high in the subterranean environment where predation pressures and climate-induced mortality will be less than at the surface. Further, anéciques do not, as a rule, reproduce by parthenogenesis (JAENIKE and SELANDER, 1979) which is a method of channelling a considerable portion of the available energy into reproduction to ensure the continued existence of the species. This is a tactic adopted by épigés that can be identified with *r* selection. Other features of such forms that represent the investment of energy in production of new individuals include small body size, developmental times of less than 1 yr (EDWARDS and LOFTY, 1977) and high metabolic rates. These, to counteract the high mortality risks to which these surface-dwellers are subjected.

8.6   Concluding remarks

The ideas embodied in *r* and *K* selection are not without their critics and it would be unwise to apply them too rigorously to the life styles of earthworms.

The distinction between surface-dwellers and deep-burrowers is not absolute, for example. Many species change their vertical distribution patterns with the season, migrating from surface layers to deeper horizons to escape summer drought. Tilling of arable soil also allows surface-dwellers to penetrate to greater depths than is possible in compacted soils. When this happens, a life style that offers some flexibility could be a distinct advantage.

This flexibility has been incorporated in various aspects of earthworm biology reviewed in the previous chapters. There is, for example, a considerable amount of overlap between épigés and endogés as far as dietary preferences are concerned. Again, the ability of épigés, enodgés and anéciques to switch from ammonotelism to ureotelism when starved suggests some measure of control over the internal environment in all three of these ecological groupings. Facultative diapause and facultative parthenogenesis, expressed to differing degrees in different species are further examples of this flexibility of life style.

References

ANDREWARTHA, H. G. and BIRCH, L. C. (1954). *The Distribution and Abundance of Animals*. Chicago University Press, Chicago.

ATLAVINIYTE, O. and DACIULYTE, J. (1969). The effect of earthworms on the accumulation of vitamin B_{12} in soil. *Pedobiologia*, **9**, 165–70.

BAHL, K. N. (1927). On the reproductive processes of earthworms: Part 1. The process of copulation and exchange of sperm in *Eutyphoeus waltoni*. *Quarterly Journal of Microscopical Science*, **71**, 479–502.

BOUCHÉ, M. (1977). Strategies lombriciennes. *Ecological Bulletin (Stockholm)*, **25**, 122–32.

BOYD, J. M. (1958). The ecology of earthworms in a cattle-grazed machair in Tiree, Argyll. *Journal of Animal Ecology*, **27**, 147–57.

CLARK, R. B. (1978). Composition and Relationships. In *Physiology of Annelids*, Mill, P. J. (Ed.). Academic Press, London and New York.

CONTRERAS, E. (1980). Studies on the intestinal actinomycete flora of *Eisenia lucens* (Annelida, Oligochaeta). *Pedobiologia*, **20**, 411–16.

DARWIN, C. (1881). *The Formation of Vegetable Mould through the Action of Worms, with Observations on their Habits*. Murray, London.

DASH, M. C. and SENAPATI, B. K. (1982). Environmental regulation of Oligochaete reproduction in tropical pastures. *Pedobiologia*, **23**, 270–71.

EDWARDS, C. A. and LOFTY, J. R. (1977). *Biology of Earthworms*, 2nd edition. Chapman and Hall, London.

EVANS, A. C. and GUILD, W. I. McL. (1948). Studies on the relationships between earthworms and soil fertility. IV. On the life cycles on some British Lumbricidae. *Annals of Applied Biology*, **35**, 471–84.

GANSEN, P. S. van (1962). Structures et fonctions du tube digestif du lombricien *Eisenia foetida* (Sav.). *Annales de la Société royale Zoologique de Belgique*, **93**, 1–120.

GARDINER, M. S. (1972). *Biology of the Invertebrates*. McGraw-Hill, New York.

GERARD, B. M. (1964). Synopses of the British Fauna, (6) Lumbricidae. Linnean Society, London.

GERARD, B. M. (1967). Factors affecting earthworms in pastures. *Journal of Animal Ecology*, **36**, 235–52.

JAENIKE, J. and SELANDER, R. K. (1979). Evolution and ecology of parthenogenesis in earthworms. *American Zoologist*, **19**, 729–37.

JAMIESON, B. G. M. (1970). A review of the Megascolecid earthworm genera (Oligochaeta) of Australia. Part II. The subfamilies Ocnerodrilinae and Acanthodrilinae. *Proceedings of the Royal Society, Queensland*, **82**, 95–108.

JOHANSEN, K. and MARTIN, A. W. (1966). Circulation in the giant earthworm, *Glossoscolex giganteus*. *Journal of Experimental Biology*, **45**, 165–72.

LAVERACK, M. S. (1961). Tactile and chemical perception in earthworms. II. Responses to acid pH solutions. *Comparative Biochemistry and Physiology*, **2**, 22–34.

LAVERACK, M. S. (1963). *The Physiology of Earthworms. International Series of Monographs on Pure and Applied Biology*. Vol. 15. MacMillan, New York.

LJUNGSTRÖM, P. O. (1972). Taxonomical and ecological notes on the earthworm genus *Udeina*, and a requiem for the South African acanthodrilines. *Pedobiologia*, **12**, 100–10.

LJUNGSTRÖM, P. O. and REINECKE, A. J. (1969). Ecology and natural history of the microchaetid earthworms of South Africa. *Pedobiologia*, **9**, 152–7.

MADGE, D. S. (1969). Field and laboratory studies on the activities of two species of tropical earthworms. *Pedobiologia*, **9**, 188–214.

MARIALIGETI, K. (1979). On the community structure of the gut microbiota of *Eisenia lucens* (Annelida, Oligochaeta). *Pedobiologia*, **19**, 213–20.

MEGLITSCH, P. A. (1967). *Invertebrate Zoology*. Oxford University Press, New York.

MICHEL, C. and DEVILLEZ, E. J. (1978). Digestion. In *Physiology of Annelids*, Mill, P. J. (Ed.). Academic Press, London and New York.

NEVO, E. (1979). Adaptive convergence and divergence of subterranean mammals. *Annual Review of Ecology and Systematics*, **10**, 269–308.

PHILLIPSON, J. and BOLTON, P. J. (1977). Growth and cocoon production by *Allolobophora rosea* (Oligochaeta: Lumbricidae). *Pedobiologia*, **17**, 70–82.

PIANKA, E. R. (1970). On 'r' and 'K' selection. *American Naturalist*, **104**, 592–97.

PIEARCE, T. G. (1972). The calcium relations of selected Lumbricidae. *Journal of Animal Ecology*, **41**, 167–88.

PIEARCE, T. G. (1978). Gut contents of some lumbricid earthworms. *Pedobiologia*, **18**, 153–7.

POKARZHEVSKIJ, A. D. and TITISHEVA, N. G. (1982). Population dynamics of the earthworm *Eisenia nordenskioldi* in meadow steppe habitats in the USSR. *Pedobiologia*, **23**, 266–7.

RAMSAY, J. A. (1949). Osmotic relations of worms. *Journal of Experimental Biology*, **26**, 65–75.

RHEE, J. A. van (1969). Development of earthworm populations in polder soils. *Pedobiologia*, **9**, 133–40.

SATCHELL, J. E. (1955). Some aspects of earthworm ecology. In *Soil Zoology*, Kevan, D. K. McE. (Ed.). Butterworths, London.

SATCHELL, J. E. (1967). Lumbricidae. In *Soil Biology*, Burges, A. and Raw, F. (Eds). Academic Press, London and New York.

SIMS, R. W. (1964). Internal fertilization and the functional relationship of the female and the spermathecal systems in new earthworms from Ghana (Eudrilidae: Oligochaeta). *Proceedings of the Zoological Society of London*, **143**, 587–608.